TVT

Drilled Pier Foundations

Drilled Pier Foundations

Richard J. Woodward, Jr.

Chairman of the Board, Woodward-Clyde Consultants
San Francisco, California

William S. Gardner

President, Woodward-Gardner & Associates, Inc.
Philadelphia, Pennsylvania

David M. Greer

Consulting Geotechnical Engineer,
Fayetteville, Arkansas

DAVID M. GREER, Editor

McGRAW-HILL BOOK COMPANY

New York   St. Louis   San Francisco   Düsseldorf   Johannesburg
Kuala Lumpur   London   Mexico   Montreal   New Delhi
Panama   Rio de Janeiro   Singapore   Sydney   Toronto

Library of Congress Cataloging in Publication Data

Woodward, Richard J.
  Drilled pier foundations.

  Includes bibliographical references.
  1. Piling (Civil engineering)   I. Gardner,
William S., joint author.   II. Greer, David M., joint author.
III. Title.
TA780.W65   624'.154   72-3984
ISBN 0-07-071783-4

1 2 3 4 5 6 7 8 9 0   KPKP   7 6 5 4 3 2

*The editors for this book were William G. Salo, Jr., and
Robert E. Curtis, the designer was Naomi Auerbach, and its
production was supervised by George E. Oechsner.   It was
set in Caledonia by The Maple Press Company.*

It was printed and bound by The Kingsport Press.

Contents

Preface

In 1963, as part of its professional development program, the predecessor firm of Woodward-Clyde Consultants held an in-house seminar on drilled pier foundations. The transcript of the meeting was converted into a set of notes for our own use. In 1968, upon David M. Greer's retirement from the firm, he was commissioned by the Woodward-Clyde Consultants' Professional Development Committee to serve as editor for a book on drilled piers. Beginning with the 1963 notes and utilizing the considerable experience of the firm with drilled pier foundations, a start was then made to compile a book which would reflect a reasonably complete treatment of current investigation, design construction, and engineering supervision practices.

This book has been written not only for the geotechnical engineer but also for the architect, structural engineer, engineering technician, contractor, builder and developer, and, in some regards, the project owner. It is our opinion that designers often can use more knowledge of how the actual work is done in the field and the tools that are used, as well as knowledge of current trends and limitations and analytical treatment of soil foundation interaction. We also believe that contractors should know more about the engineer's concept of engineering supervision and because builders and developers are often expected to make decisions involving project time and costs that they too would be aided by having some background knowledge of drilled pier foundations.

Much of the material used in this book was obtained through the experience of individual engineers within the affiliated firms of Woodward-Clyde Consultants. The writers are particularly indebted to T. A. Bellatty, R. A. Millet, and D. C. Moorhouse for their contributions to Chapters 5 and 7 and to J. K. Mitchell and S. T. Thorfinnson for a constructive review of the text. In addition, interviews were conducted with contractors, equipment manufacturers, consulting engineers, drillers and inspectors, as well as teachers and researchers in the field. In Appendix C, we have listed many of the organizations that were consulted or provided material used in the book. In addition, we want to thank the many other people who gave us counsel whom we have failed to mention. Finally, particular thanks are due the Professional Development Committee of Woodward-Clyde Consultants whose generous sponsorship made this book possible.

RICHARD J. WOODWARD, JR.

WILLIAM S. GARDNER

DAVID M. GREER

Drilled Pier Foundations

CHAPTER ONE

State of the Art

1-1 SCOPE OF THE BOOK

This book is principally about the foundation element variously called a "drilled pier," a "drilled caisson," or a "large-diameter bored pile" (British usage). This type of foundation element when made with an enlarged base is variously referred to as a "drilled and underreamed caisson," a "belled caisson," or a "belled pier." When made without an enlarged base, it is called a "straight-shaft caisson" or "straight-shaft pier" or an "unbelled pier," and when such a pier is drilled into rock and designed to transfer load by sidewall shearing resistance, as well as by end-bearing, the term "socketed caisson" or "rock socket" is often used.

In this book we shall use the terms "drilled pier" and "drilled and underreamed pier" or "drilled and socketed pier" for these foundation elements. An enlarged base will be referred to as an "underream" or as a "bell," interchangeably.

There are other uses of large-diameter drilled holes in underground construction, in addition to installation of the foundation elements mentioned above, which are important enough to merit consideration in a book of this sort. We are therefore including, where the authors' experience indicates that the application may be of importance, discussion of some additional structural elements or construction techniques using the same drilling machines as are used for pier holes. Some of these are:

Very large diameter drilled holes for access shafts, missile silos, mineral exploration, etc.

Slanted or battered holes for tieback anchors, drainage, or other uses (including batter piles)

Retaining walls and bracing for excavations (both tieback and cantilevered types)

Slurry trench construction (for seepage cutoff or for load-bearing walls)

Landslide stabilization piers and drains

This is a "state-of-the-art" book. The art is changing. New techniques, new and evolving machines, and increasing skill on the part of the specialty contractors who do most of the work, all have led to many new capabilities in the field in the past decade, and will undoubtedly result in significant new developments and capacities in the next decade.

This is not a theoretical treatise, and the chapter "Design Considerations" (Chap. 3) is not intended as a design manual. Instead the authors have attempted to select from their experience—influenced by wide variations in geography, geology, personality, and environment—significant observations and lessons that they have gathered and applied in their own work, and to assemble from the literature scattered but relevant work relating to the way loads are really transferred to soil or rock, as distinguished from the much-simplified assumptions usually used in design. They have tried to present these observations and lessons in a way that will help the interested reader understand and anticipate many of the problems that he may encounter in the use of large-diameter bored holes in construction. The authors hope that this understanding may help the reader avoid many of the mistakes that they have seen—including some they made themselves. Mistakes in this field often can be very expensive to all concerned. They may even result in complete disaster—physical or financial, or both.

The book does not include consideration of driven piles, drilled-in piles (augered piles), "displacement caissons," or "pressure-injected footings" (Franki piles), nor does it cover drilled or mole-excavated tunnels. It treats only of installations in large-diameter machine-drilled (or bored) holes which have had all the soil removed (and usually, but not always, all the water).

Another related type of foundation not considered in this book is the "drilled-in caisson" [1.1, 1.2],* used widely in New York City and elsewhere in the United States where very heavy loads are involved and sound rock, usually at considerable depth, is available to carry the loads. In this foundation type, a heavy steel shell, with a very heavy steel cutting shoe, is driven to rock by a pile driver and is then cleaned out by blowing, jetting, or augering. Bearing is usually attained by drilling a rock socket with a churn drill, a reinforcing beam or cage is positioned in the casing, and the entire system is then concreted. This special type of deep foundation is well known, and differs from the types of construction treated in this book in that the steel casing is very heavy and is driven to rock, rather than being installed by boring techniques.

The term "large-diameter" refers to holes larger than 15 in. (38 cm) and usually at least 24 in. (61 cm) in diameter. Some of the holes may be very much larger. Pier shafts 10 ft (3 m) in diameter and underreams 20 ft (6 m) or larger are not uncommon. An ASCE-ASA publication [1.3] defines a foundation pier as one which is at least 2 ft (61 cm) in diameter; the Canadian National Building Code defines it as a shaft with a bearing surface having a minimum diameter of about 5 ft (1.5 m).

1-2 DRILLED PIERS: ADVANTAGES AND DISADVANTAGES

The principal reason for using drilled piers in preference to other types of deep foundation is economy. In ground where soil, rock, and groundwater conditions are suitable for their use, they often cost less than other types of foundation per unit of load carried. Where soil and rock conditions are most favorable, the saving involved in the use of drilled piers can be very substantial, sometimes amounting to half the cost of the nearest competitive type of foundation. An example of such a cost advantage is a case occurring in 1951, when a refinery in New Jersey found it less expensive to have a truck-mounted pier-drilling machine sent from Los Angeles and back than to have pile foundations, conventional footings, or piers constructed by local contractors. The contract involved some very heavy column loads for refinery units, founded on shale below fill and glacial till, with the water table near the surface. There were no pier-drilling machines available on the East Coast of the United States at that time.

The potential economic advantage of drilled pier foundations comes from the fact that rotary drilling is a rapid and inexpensive way of excavating, provided (1) a circular excavation, usually with its length substantially greater than its diameter, can be adapted to the proposed

---

* Bracketed numbers refer to References at end of Chapter.

structural requirements; (2) the material to be excavated can be cut readily with rotary drilling tools; and (3) the soil does not cave or flow into the hole. As any one of these advantageous conditions diminishes, the cost of drilled pier construction goes up. As will be shown later, however, there are often ways of circumventing difficulties due to ground conditions; and even in the absence of ideal conditions, drilled piers often retain an effective cost advantage.

In addition to cost comparison, the following features, favorable and unfavorable, are among those considerations which may influence the choice between drilled piers and other types of foundation:

Favorable Factors

Early Foundation Construction Is Facilitated.   Drilled piers can often be completed before grading operations and basement excavation are undertaken, appreciably expediting overall completion of the job.

Rapid Completion Is an Advantage.   When surface and subsurface conditions are suitable, the rapid completion of piers not only can expedite the foundation phase of the project, but also can expedite subsequent operations because of minimal interference with other construction operations. (This advantage is subject to the effects of weather and of unexpected ground conditions, and as a result pile foundations, though nominally more costly, may be a better choice for some jobs.)

A Single Pier Often Can Replace a Pile Cluster.   The result is elimination of pile caps and related form work.

Piers Can Be Drilled through Cobbles and Small Boulders.   Suitable machinery will drill through stones which would deflect piles, and so redesign is eliminated. (Redesign of more than 30 percent of the pile caps is not unusual for projects involving piles driven through bouldery soils or fill.)

Reinforcing and Form Construction Are Minimized or Eliminated. This produces savings when compared with conventional excavation and footing construction.

Pile Hammer Noise Is Eliminated.   Most rotary-type pier-drilling machines are relatively quiet; however, machines using percussion or compressed air, for drilling through rock, may be very noisy.

Vibration, Heave, and Ground Displacement Are Eliminated.   This is a distinct advantage when nearby structures or machinery must not be disturbed. When displacement piles are driven into saturated firm-to-stiff clays, ground heave and lateral displacement of previously driven piles may become a major problem.

Uplift Resistance Is Easily Provided.   Such resistance is readily developed in ground where underreams (bells) can be formed or where the sides of the shaft in the bearing stratum have enough contact to assure the development of shearing resistance between shaft and bearing material.

Uplift or Downdrag Can Be Reduced. Uplift loads due to swelling soil, as well as downdrag loads due to settling soil ("negative skin friction"), can be avoided or minimized more readily than with most other types of deep foundation because drilled piers can be double-cased easily.

Bearing Surfaces Are Exposed for Inspection. A drilled pier hole, if it can be completed at all, can usually be inspected to make certain that it has reached suitable bearing. This is not true of a driven pile, and there will be many circumstances under which the amount or condition of the piles' contact with the proposed load-bearing formation remains uncertain. Under some geologic circumstances, this is a definite advantage for the drilled pier type of foundation.

Unfavorable Factors

Operations Are Affected by Bad Weather. Wet weather interferes with pier drilling and concreting to a much greater extent than with pile driving.

Unfavorable Soil Conditions May Interfere. Such conditions, which may be unexpected, may introduce major difficulties and delays in drilled pier construction. This is true, of course, of any foundation system, but drilled piers are (in the authors' opinion) more sensitive to this element than are competitive types of deep foundation.

More Complete Soil Exploration Is Needed. For drilled pier applications the investigation needs to be (again, in the authors' opinion) more thorough than for most other types of deep foundation.

Building Codes and Governmental Regulations May Be Unfavorable. In some cases these are so formulated that the full load capacity cannot be used, or they impose construction restrictions that increase drilled pier costs materially.

Inspection and Technical Supervision Are Critical. These are more critical for pier drilling, and especially for the concreting operation, than for most other deep-foundation systems. (Cast-in-place concrete piles also are critical in this respect.)

Ground Loss May Be Substantial. Excess excavation and consequent settlement of adjacent structures are possible under some circumstances, and their prevention requires careful attention in both design and construction.

These characteristics, and their importance, are discussed at more length at appropriate places in the chapters that follow.

1-3 DEVELOPMENT OF LARGE-HOLE DRILLING MACHINERY

The use of deep hand-excavated piers for foundations is probably as old as civilization. Yet up to about 30 years ago, the holes for pier foundations were still, for the most part, hand-excavated. The two most common procedures in the United States were those described in the textbooks under the subjects "Gow caissons" and "Chicago caissons." Undoubtedly a few machines for excavating pier shafts were improvised in areas where the soils were particularly suited to machine drilling, some of them going back to the days of steam power.

The earliest record the authors have found of the use of machine-drilled piers for foundations comes from San Antonio, Texas. Mr. Willard Simpson, Sr., a consulting engineer, recognized early in the 1920s that buildings in that area could be protected from the very destructive effects of soil swelling and shrinkage only if all foundations were supported at levels below the zone of seasonal moisture variations (which in San Antonio might be 25 ft or more deep). After experimenting with a few pier foundations excavated manually with the long-handled shovels that utility companies used for making pole holes, Mr. Simpson and Mr. Ed Duderstadt, a local contractor, adapted a well-drilling machine operated by four men pushing a capstan bar around a circular track (a machine illustrated in Dempster Mill Manufacturing Company's 1895 catalog). This worked so well in San Antonio soils that the contractor was soon operating six or eight of these machines, horse-powered instead of man-powered (Fig. 1-1). In about 1938 one of these machines was mechanized by combining it with the power unit of a steam shovel. At least one of the horse-powered machines was still in use in 1951 [1.4].

Mr. Eric P. Johnsen of Abbott-Merkt & Company relates that drilled piers with 30° bells were used under the Continental Can Company Building in the (then) Union Pacific Industrial Area in Los Angeles in 1928. The drilling machine was mounted on a truck; further details are lacking.

During the 1920s the Gow Company of New York City built and used a bucket-type auger machine, electrically powered and mounted on the turntable frame of the crawler tractor of a crane. The drill bucket, about four ft in diameter, was equipped with adjustable "reaming knives" that enlarged the shaft hole to diameters up to at least 7 ft. Bells with 30° sides (angle measured from the vertical) were formed at the bottom by hand labor, using air spades for digging and electric winches for hoisting. One pier, extending to a depth of about 120 ft, was completed (including concreting) in about 10 hr [1.5, 1.6].

Fig. 1-1 An early horse-powered bucket-auger well-drilling machine, from Dempster Mill Mfg. Co. 1906 catalog. (*Dempster Industries, Inc., Beatrice, Nebr.*)

The development of efficient, reliable, and commercially available machines of this sort for widespread general use did not come until after World War II. During the war years, the pressure for rapid construction of many light buildings for the armed services resulted in widespread use of the truck-mounted auger machines which had been in use for a few years by the utility companies. These "pole-hole machines" were used to drill uncased holes for small pier foundations, usually shallow and not more than 18 in. (46 cm) in diameter. The work was scheduled so that the concrete truck followed immediately after the drilling machine. No forming was necessary, and the work went very rapidly. The result of the demand was the development of many small pier-drilling contractors, the most enterprising of whom started seeking other and larger work and devising their own drilling tools and, in some cases, drilling machines. In particular, tools were devised for making an expanded base (an underream or bell) without enlarging the shaft, so that larger column loads could be carried without wasting concrete on unnecessarily large-diameter shafts. Most of the machines operating in 1946 were custom-made, each model being built to match the experience of the builder.

The greatest impetus to this development occurred in two states: Texas and California. Because of differences in geology and in the soils and the way they handled, as well as differences in structural requirements, the development of "drilled pier machines," as they soon came to be called, diverged in these two areas. The result was the eventual evolvement and production of two substantially different kinds of auger-type pier-drilling machines: the "bucket-auger" machine in Cali-

Fig. 1-2 Calweld Model 200-B with 30-in.-diameter drilling bucket and 30-ft triple-telescoping kelly. (*CALWELD, Division of Smith Industries International, Inc., Santa Fe Springs, Calif.*)

Fig. 1-3 Hughes CLLDH "super duty" crane-mounted rig working on pier holes for slide stabilization on the Seattle Freeway. (*Hughes Tool Company, Inc., Houston, Tex.*)

fornia (Fig. 1-2) and the open-helix-type auger machine in Texas (Fig. 1-3). (The latter should not be confused with the much smaller "continuous-flight" auger machine that is used in soil exploration and in predrilling holes for piles.) Each type of machine has its own distinct advantages under certain favorable circumstances and its own following among contractors.

As these commercially available machines evolved, improvements were made in several directions, responding to the needs of the contractors who used them.

1. Larger machines were designed, leading each manufacturer to develop a series of models of different sizes and capabilities.

2. The larger machines were designed to apply higher torque to the cutting tools, enabling the machines to drill harder rock, and to turn larger-diameter soil-drilling tools, than had been possible before.

3. Telescoping "kellys" (a square or splined shaft that can slide up and down through the rotary driving head while being rotated) were

developed and increased in length and cross section, so that deeper holes could be drilled without coupling and uncoupling the drill stem.

4. Some light models which had initially depended on the dead weight of kelly and auger for downward force on the cutting edges were eventually equipped with hydraulic "pulldown" or "crowd" mechanisms to produce faster penetration into hard formations.

5. Speed in setting up was greatly increased by the addition of hydraulic raising and lowering of the mast, hydraulic outrigger jacks for leveling, and in-and-out slide and rotating base for positioning the auger over the exact point for boring.

6. Crane mountings were devised for some of the larger machines (instead of the usual truck mounting), allowing the handling of longer and heavier kellys, as well as larger augers and underreaming (belling) tools, and the placing and pulling of longer and heavier casing sections (Fig. 1-4).

7. Cutting bits for drilling rock were greatly improved, so that the higher-torque machines could drill further into harder rock than before,

Fig. 1-4 Calweld Model 150-CH crane attachment with 36-in. heavy-duty 30° belling bucket. (*Caisson Corporation, Chicago, Ill.*)

Fig. 1-5 Modified oil well drilling rig, for drilling a 10-ft-diameter hole more than 5,000 ft deep. (*Loffland Brothers Company, Contractors, Tulsa, Okla.*)

still using their auger-type tools or specially adapted rock-drilling attachments.

8. Oil well drilling tools and techniques were adapted to large-diameter hole drilling (Fig. 1-5). For example, drilling mud was used to prevent the caving of water-bearing formations, permitting rapid drilling of the shaft hole (to be cased off later) down to the bearing or belling stratum.

9. Entirely new machines were put on the market, such as the Benoto "hammer-grab" machine (made in Europe) and the Ingersoll-Rand Magnum drill, which uses an assembly of pneumatically driven chopping bits and blows the cuttings out of the hole with compressed air.

1-4   DEVELOPMENT OF THE DRILLED PIER

In the chapters that follow, references will be made to four types of pier, which, while similar in construction technique, differ in the way in which they are assumed, for design purposes, to transfer the foundation load to the earth. These types are illustrated in Fig. 1-6.

Straight-shaft end-bearing piers develop their support from end-bearing on strong soil, "hardpan," or rock. The overlying soil is assumed to contribute nothing to the support of the load imposed on the pier.

Straight-shaft sidewall shear, or "friction," piers pass through overbur-

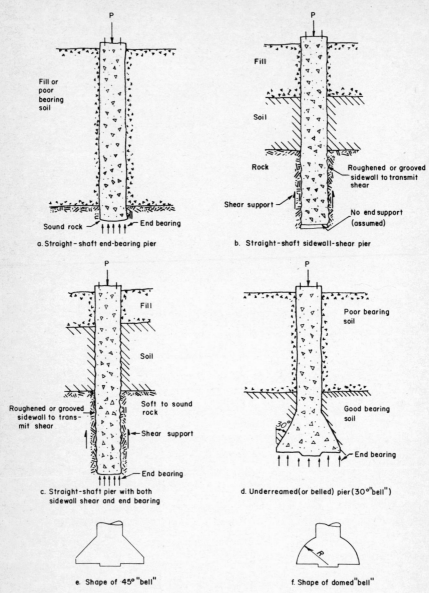

Fig. 1-6 Types of drilled piers and underream shapes.

den soils that are assumed to carry none of the load, and penetrate far enough into an assigned bearing stratum to develop design load capacity by sidewall shear between pier and bearing stratum. (These are sometimes called "friction piers"; however, we prefer not to use this term because in many cases the support is developed in shearing strength of cohesive materials rather than either internal friction (ϕ) of soil or adhesion between soil and pier.)

Combination straight-shaft, sidewall shear and end-bearing piers are of the same construction as the two just described, but with both sidewall shear and end-bearing assigned a role in carrying the design load. When carried into rock, this pier may be referred to as a "socketed pier" or a "drilled pier with rock socket."

Belled or underreamed piers are piers with a bottom bell or underream. In United States practice, all the load bearing capacity is always assigned to the bell in end-bearing.

In the decade following World War II, in areas where ground conditions were especially suitable for the use of rotary pier-drilling machines, the economy of drilled piers over all other types of deep-foundation construction was immediately evident, and they quickly became the prevalent type. In many cities—for example, Houston, Denver, San Antonio—pile foundations continued to be used only in exceptional circumstances, usually in locations involving caving soils and great depths to suitable bearing strata.

As the use of the drilled pier became common, some designers and some contractors lost sight of the limits of the soil, groundwater, and rock conditions which had led to the advantages of drilled piers. The resulting attempts to use them where the geologic formations were not suitable for the drilling machines, tools, and techniques then available sometimes led to unsatisfactory construction performance in the form of delays, "extras," and occasional abandonment of contracts, with the job being redesigned for piling or some other form of foundation. The difficulties developed most commonly under the following circumstances: in caving soils (especially cohesionless soils below the water table), where casing had to be used, and the available machines were not suitable for placing or pulling the necessary sizes of casing; in bouldery soils, where sometimes stones were encountered which could neither be drilled nor removed intact; and in soil-rock areas, where drilling was stopped by ledges or pinnacles which had to be broken up by jackhammer before the pier hole could be completed to the planned bearing level.

These difficulties, in turn, led to the development of specialized tools, new machines, and better techniques, thus expanding the range of economical application of drilled piers. As a result, drilled piers are now used in many areas and on many jobs where ten years ago their use would have been uneconomical or actually unfeasible. For example, where water-bearing formations have to be penetrated and cased off, the larger and more powerful machines can often enable the contractor to drill-in the casing ahead of the excavation, or to drill the hole quickly and place the casing before substantial caving has occurred. Most modern pier-drilling machines also can be adapted for pulling the casing smoothly and at a controlled rate when concrete is being placed. With the larger machines, casings can also be drilled into rock to seal off water inflow, and many contractors' experience in sealing off casing-rock contacts has led to skill and speed in this operation that were generally lacking a few years ago. The accumulation of specialized experience on the part of contractors has led to their being able to handle, sometimes as a routine operation, difficulties which would have slowed the job greatly ten years ago.

In addition, there has been a buildup of skilled machine operators on the contractors' work forces. This constitutes a major change in the industry. The drilling of large-diameter holes in the earth requires a very special skill, and to do it without trouble or delay under potentially troublesome conditions makes the difference between profit and loss for the contractor, and may mean a job finished on time and without controversy, claims, and confusion, as far as the owner and his agents are concerned. A substantial part of the "art" of drilled pier foundation construction lies in the personal skill of the drilling-machine operator and the experience and ingenuity of the operator and the drilling contractor's superintendent.

1-5   DRILLED PIERS NOW IN WIDESPREAD USE

The local occurrence of clearly favorable ground conditions, plus the development of machinery, techniques, and skills for coping with the difficulties that arise under less favorable conditions, have led to the use— or at least the trial—of machine-drilled piers in all parts of the world. In many areas they have become the dominant type of foundation. At the time of writing (1971) there are pier-drilling contractors, either locally available or on call from a city not too far away, in any part of the United States. Drilled piers have been adopted with enthusiasm in England, where there is extensive current research into their bearing capacity, and their use is spreading rapidly through Canada and Latin America. Drilling-machine manufacturers are now exporting their products to all continents.

Despite the widespread availability of drilled piers and their virtual monopoly of the market for economic reasons in some areas, they are still not well known in all areas. As a result, a major building is occasionally designed to be supported on piles in an area where piles cannot compete economically with drilled piers. An example is a building in Houston, Texas, where the pile foundations shown in the plans would have cost approximately $100,000 more than did the drilled piers that were used after the foundations had been redesigned. The building was originally designed in another city, in a region where pile foundations are common and drilled piers are rarely suitable.

1-6   BATTER PIERS AND ANCHOR PIERS OR TIEBACKS

With the early pier-drilling machines, sloping holes could be drilled only by blocking up one side (or end) of the truck on which the drilling machine was mounted. This usually limited the slope to not more than 1 horizontal to 4 to 6 vertical. The later machines, however, have been conveniently arranged to drill with the mast and drilling assembly tilted back over the cab (in the case of truck-mounted machines), or forward (in the case of crane-mounted rigs). As a result, where suitable equipment is available, pier holes can now be drilled with batters as flat as 1:1. Batters flatter than 1 horizontal to 3 vertical, however, are difficult to keep in alignment, and are limited in economical application to the most favorable ground conditions. Pier holes with any batter at all are more expensive than vertical ones. This places a limit on their economical use, rather than on what can be done if necessary. A further limitation that must be kept in mind is that, even with ground conditions that are ideal for forming underreams in vertical pier holes, an expanded base is much more difficult to form at the bottom of a sloping hole, and will be much more expensive and less certain as to ultimate shape, dimensions, and the completeness and continuity of concreting.

Smaller-diameter holes for anchor piles or tiebacks, however, are readily drilled with specialized drilling machines. With suitable equipment they can be drilled at practically any angle, including slopes upward from the drilling point (as needed for drainage installations, for example). Because the dimensions and shapes of enlarged bases (or ends) for tieback anchor holes are not as critical as are those of the underreamed bases for piers, the disadvantage of uncertainty of base shape and dimensions mentioned above is here minimized.

1-7   DRILLED PIERS IN CAVING GROUND: SOME SPECIAL TECHNIQUES

Except for rock, the most common impediment to large-hole drilling is the need to pass through (and case off) water-bearing formations. In the early years of the industry, this usually was not attempted. As the uses of large-diameter drilled holes spread—both geographically and geologically—many jobs were found to be feasible with the aid of conventional dewatering (by well points, deep wells, etc.). In these applications, the speed of completion of drilled holes, as compared with conventional excavation and forming, greatly reduced the time that dewatering had to be maintained, thus effecting an additional substantial saving. The use of minimal dewatering installations often permits pier holes to be drilled and promptly filled with concrete in ground where much more elaborate dewatering installations would have been required if more conventional excavation and construction techniques had been used. Well points have been employed for temporary stabilization of fine-to-medium sand where capillary forces in the dewatered but moist sand were sufficient to keep the hole from caving for a few minutes. Deep wells have been used successfully for stabilizing clay with sand and gravel layers which otherwise would have produced very troublesome sloughing and caving; and in fissured clays where the deep wells, pumped down to a level 10 ft below the bottoms of the footings, allowed 8-ft-diameter underreams to be cut and concreted in the dry and without caving. Dewatering techniques thus can be used occasionally to convert a potentially troublesome operation into a near-routine one.

The use of drilling mud to prevent caving, a technique adapted from oil well practice, has been used successfully to extend the range of application of drilled piers into ground conditions that would otherwise not have permitted either the drilling of an open hole or the economical use of casing alone. The technique is often not understood by people who have not had experience with it. However, it has an important potential for economy when conditions are otherwise suitable for carrying the foundation loads on drilled piers (though unfavorable for their construction by ordinary means) and, for one reason or another, unsuitable for the installation of other types of deep foundation. An example is an addition to a paper mill, involving foundations to support heavy loads, in a location where no disturbance to existing machinery could be tolerated and where 20 or more feet of loose water-bearing sand overlie the bearing stratum of stiff clay. The pier shafts were drilled down to the clay in mud-filled holes. The casing was then inserted and sealed into the clay. The mud was removed from the casing, and the underream was then cut in a dry hole with the usual underreaming tool. The pier was completed without difficulty. All of this was accomplished without vibration, whereas pile driving would almost certainly have disturbed the adjacent machinery in the plant.

The use of drilling mud in drilled pier construction has usually been limited to keeping a straight shaft from caving. There have been several instances of shafts *and underreams* being drilled entirely in mud-filled holes, however, with the concreting of the completed footing and shaft being accomplished entirely by later displacement of the mud, from the bottom up, with pumped-in or tremied concrete.

Another innovation in handling the problem of casing off water-bearing sand or gravel is to install the casing with a "vibrodriver," which vibrates the casing through the troublesome stratum and seats it into an underlying impervious stratum. This method offers the advantages of speed; easy, effective seating of the casing into the bearing stratum; and maximum lateral support for the pier.

1-8 THE STATUS OF DESIGN

At this time, the design of drilled pier foundations (as far as soil-pier or rock-pier interaction is concerned) remains mostly empirical. The principles of soil and rock mechanics are often applied, as a routine procedure, to the determination of "safe bearing values" and the prediction of performance of soil- or rock-bearing piers. But this approach suffers from incomplete development of theory for deep-based foundations. It is limited also by the difficulty of obtaining representative "undisturbed" samples of the materials which are expected to carry the foundation loads and the difficulty of simulating in the laboratory the *in-situ* environment and the imposed stress conditions. Laboratory testing to provide data on the strength and compressibility of deep bearing strata has not been neglected. However, such testing is difficult under ideal conditions, and virtually impossible when dealing with such materials as hard (but fissured or slickensided) clays; "hardpan"; stony glacial till; and layered, fissured, or weathered rock. Design criteria based solely on laboratory tests on such materials either may be unrealistically conservative or may be based on a shearing strength that is unrealistically high.

Commonly, where piers are carried into rock or where sidewall shear is relied on for pier support, the design approach is almost entirely empirical, occasionally being predicated on the results of field loading tests. Recent advances in the relatively new art of rock mechanics promise to permit a more rational organization of experience and, eventually, to make possible a less empirical approach to the design of rock-based piers.

Recent computer-stimulated advances in analytic techniques (such as the finite element method), applied to the study of elastic and visco-elastic behavior of "ideal materials" under stress, promise an improved understanding of the distribution of stresses around and beneath pier (and pile) foundations, both in rock and in soil.

At the present time, however, the known uncertainties in theory and the questionable value of results of laboratory tests on some materials have led to reliance on large-scale or even full-scale loading tests for development of design parameters for many sites. Many full-scale load tests are run each year, most of them to check safe bearing for a specific load and pier type in a specific rock or soil formation. Some of these test results are reported in the literature, but most of them are not.

Full-scale load tests are expensive, and the variety of sizes, shapes, depths, soil and rock conditions, and loading objectives is practically infinite. The results of most load tests are therefore not applicable, with a reasonable degree of confidence, to other designs and geologic circumstances. (The same thing is true, of course, for pile foundations or any other deep-foundation system.)

The result of these difficulties and limitations has been that the recommendations of the geotechnical engineer–geologist team that makes and reports foundation investigations and recommends design criteria for drilled pier foundations have usually been conservative. Allowable unit loads have increased slowly as experience both with full-scale load tests and with completed structures has accumulated. The gradual increase in proven allowable unit loads has been reflected, in some cities, by increases in building code allowable pressures, but usually the unit loads on rock which are justified by experience are substantially higher than those allowed by the building codes.

Some of the research that has been done on the distribution of loads between end-bearing and sidewall shear has clearly indicated that more of the load is carried by sidewall shear than is usually assumed. Many building codes will allow either end-bearing or sidewall shear—but not both—to be used in design, and most drilled piers are therefore designed as though only one mode of support were effective. Undoubtedly such designs are sometimes unnecessarily conservative. As our ability to calculate the distribution of loads between sidewall shearing resistance and end-bearing improves, more economical designs will result.

Despite all prospects of improved understanding, it does not seem likely that, in the near future, there can be a completely analytical approach to the prediction of the interaction between soil or rock materials and deep foundations. As the art stands today, all available knowledge and techniques (analytical, experimental, and experience-based) are needed, and still the end result depends to a great extent on the judgment of the person making the analysis.

Some of the empirical and semiempirical methods that have been developed for determining design parameters for drilled piers in soil and in rock are presented in Chap. 3, "Design Considerations."

REFERENCES

1.1 White, R. E., Development in Large-diameter Piles or Caissons in the United States, *Proc. 3rd Asian Reg. Conf. SM & FE**, *Haifa, Israel*, pp. 79–85, 1967.

1.2 White, R. E., Heavy Foundations Drilled into Rock, *Civil Eng.*, vol. 13, no. 1, pp. 19–32, January 1943.

1.3 *American Standard Building Code Requirements for Excavations and Foundations*, ASCE Manual no. 32, ASA A56.1 pp. 15–16, 1952.

1.4 Old Dobbin Comes Back, *San Antonio Light*, Dec. 3, 1951.

1.5 Christie, H. A., Boring Machine Digs Wells for Concrete Piers, *ENR†*, pp. 105–106, July 28, 1932.

1.6 Beretta, J. W., Drilled, Reamed Holes for Bridge Foundation Piles, *ENR†*, pp. 217–218, Feb. 11, 1937.

* *SM & FE* = Soil Mechanics & Foundation Engineering.

† *ENR* = *Engineering News Record*.

CHAPTER TWO

Considerations Leading to the Selection of Drilled Pier Foundations

2-1 CHOICE OF FOUNDATION TYPE

The most common reason for selecting a drilled pier foundation in preference to some other type is that—for the specific structure and site conditions involved—drilled piers will cost less than the other type of foundation. In some circumstances the advantage may be less direct: for example, lack of vibration or transmitted shock during construction, the facility with which provision may be made for expansive soils or rock, or availability of equipment and materials. However, as is discussed in several places in this book (see especially Chaps. 4 and 8), the economical construction of drilled piers is sometimes more sensitive to variations in ground, groundwater, and weather conditions than that of competitive foundation systems (driven piles, for example). And sometimes site conditions are definitely unfavorable to the economical use of drilled piers, even though their construction is quite feasible. We point out in this chapter some of the circumstances which are favorable and some that are definitely unfavorable (or, at least, offer construction difficulties which must be taken into account), and we discuss the basis for decision when both favorable and unfavorable factors are present—i.e., "borderline" conditions.

20

Another choice that may need careful consideration is which type of drilled pier is most suitable. End-bearing on rock? Rock socket? Straight-shaft or underreamed? (See Fig. 1-6.) Considerations which affect these choices are discussed briefly in this chapter, and are treated also in Chap. 3, "Design Considerations."

It might be pertinent to mention at this point that the choice of foundation type is sometimes so well established by local experience and custom that there is no problem of decision. For example, in Houston, Texas, and in Denver, Colorado, geologic conditions are generally so favorable, equipment is so readily available, and pier-drilling contractors are so numerous that pile foundations are not even considered unless there are exceptional circumstances (e.g., waterfront construction). But local custom is not necessarily the best guide. In the late 1940s, when practically all foundations constructed in Houston were drilled and underreamed piers, local custom in Beaumont, only 86 miles away and in the same geologic formation, still favored pile foundations; and it was not until some years later that the economy and satisfactory performance of drilled piers brought them into general use in Beaumont.

2-2 FAVORABLE CONDITIONS

Drillable Formations

Any soil or rock that can be readily and rapidly drilled with large-diameter augers, and in which the hole will stand open without caving or squeezing in (even if for only a few minutes), is a likely setting for drilled pier operations. When underreams (bells) are to be used, the formation in which the bells are to be formed must be capable of standing open until it can be filled with concrete.

Bearing Formation at Accessible Depth

Depending on structural requirements and local geology, the range of depth required may vary from 6 or 8 ft (about 2 m) for underreamed footings for light industrial structures in stable soil areas to 150 ft (46 m) or more where piers supporting heavy loads must develop their load-bearing capacity by sidewall shear ("friction" piers) or where rock suitable for support of heavy column loads is very deep (as in the Chicago area, for example). Whenever suitable bearing formations are within the depth range of available equipment, drilled piers can be considered.

Ready Surface Access

A site that is reasonably level, has a firm surface, and provides adequate overhead clearance is a very favorable condition. Because much of the available drilling equipment is truck-mounted, good trafficability is an important consideration in the use of auger rigs and auxiliary equipment.

Suitable Equipment Available

The depth range that is commonly required to reach satisfactory bearing naturally dictates the depth range of available equipment in a given area. However, requirements that are new to an area, such as larger and deeper piers for a building of unprecedented height, may require different equipment. "Availability" should, in such cases, be discussed with prospective pier-drilling contractors. Either the exercise of contractor's ingenuity or the importation of a suitable machine from a distance may alter the "availability" picture.

Dry Hole—Freedom from Caving

Absence of subsurface water inflow is one of the most favorable conditions for speed and economy in drilled pier construction. This does not necessarily mean that the pier hole risks trouble if it extends below the water table. Many formations, particularly in clayey soils or rocks, will stand open without "bleeding" or significant water entry even though the pier hole extends many feet below the "true" (though not readily discernible) water table. Most cohesive soils will stand open above the water table without caving at least for a limited time.

Suitable Building Code Provisions

The acceptability of suitable types of drilled pier foundations, the allowance of reasonable unit loadings, and especially the proviso that exceptions to code limitations will be made for designs that are backed up by suitable tests and calculations are factors that can open up a local area to the use of drilled pier foundations, sometimes with substantial foundation cost economies.

Special Requirements

Special construction or design requirements of various kinds may dictate the choice of a drilled pier foundation over other types of foundation which might be nominally less expensive, or even structurally more suitable but for some special circumstances. Examples are (1) foundations which must be installed without transmission of vibration or shock to adjacent structures or equipment; (2) sites where heaving produced by pile driving could damage adjacent structures; (3) underpinning or new foundations within an existing structure, where access and headroom are limited, reaction for pile jacking is not available, and pier-drilling equipment (usually special machines) can operate in the available space.

Swelling Soils or Rocks

Under most circumstances, drilled piers have several advantages over other foundation systems for resisting or minimizing the structural effects of swelling soils or rocks. They provide an economical way of carrying the load-bearing element below the expansive layer, or at least below the zone of seasonal moisture change. Where the bearing surfaces can be thus taken down to a stable material, the pier shaft is more easily insulated from sidewall uplift or downdrag due to swelling or shrinking strata above than is a driven pile. And where the active formations are too deep to be economically bypassed, it is easier to resist uplift by concentrating the foundation pressure on a potentially swelling surface when the load is carried by a single pier than when it must be transmitted by a cluster of piles. (See Appendix E, tabulation for Denver area.)

Ability to Resist Uplift and Lateral Loads

Drilled pier foundations, either those with enlarged bases (bells or underreams) or straight-shaft piers, can be designed to resist uplift forces far in excess of those that can be safely applied to driven piles reaching comparable depths. The normally large shaft diameter, and the ability to readily construct a shaft with an enlarged diameter in the upper portion, as well as to underream, provides the drilled pier with a much higher resistance to lateral deflection than can be obtained with an equivalent single pile or small group of piles.

2-3   UNFAVORABLE CONDITIONS

Ground conditions or structural requirements which might have ruled out consideration of a drilled pier type of foundation in 1960 might not be of nearly so much importance today. However, these conditions will still affect costs, and should be considered as "unfavorable" factors in deciding whether or not a drilled pier foundation is the best choice for a specific project. The presence of one or two such factors might have little effect on the choice, considering the capabilities of the prospective contractors, while operation of a larger number of them might entirely rule out drilled piers. (Sometimes even a combination of several unfavorable conditions might still leave drilled piers as the preferred type—see "Special Requirements" above.)

Some of the conditions that must be weighed on the unfavorable side are listed here.

Caving or "Squeezing" Formations

Any formation, material, or condition that experience has shown to give trouble with caving, "boiling," squeezing in, sloughing, or collapse of bells or shaft walls should be considered as a possible source of trouble. Cohesionless silt, rock flour, sand, or gravel, below the water table, is particularly troublesome.   There are techniques for coping with all of them (see Chap. 4), but their presence must be expected to increase costs.

Fissured, slickensided, or jointed clays and shales are potentially troublesome, although in many instances they will stand, even in an underream, for periods varying from a few minutes to a few hours, and drilled piers are usually quite practicable if concrete placement is prompt.

Fine sands and silts *above the water table* are much less troublesome because they are often moist enough to have substantial "apparent cohesion" due to capillary forces.   These soils will often stand open in shafts, but rarely in underreams.   Coarser sands, gravels, and sand-gravel mixtures are troublesome whether wet or dry; they will cave extensively during drilling and have a tendency to "seize" on a casing that is being used to control caving.

Soft organic silts and silt-clays, "muds," and soft clays all tend to close in a bore hole, sometimes very rapidly; and sometimes lifting the auger (especially a bucket auger) will "suck in" the bore hole to almost complete closure.   Again, there are techniques for coping with this condition (use of casing or drilling mud), but they involve additional expense.   Soft or loose overburden might be favorable to the construction of pile foundations while it is precisely the opposite for drilled piers.

Boulders and Cobbles

Boulders and cobbles will obstruct and slow down pier drilling, and may make it very expensive.   Cobbles, particularly when in a clay or "hardpan" matrix, can usually be removed with ordinary drilling techniques (and suitable drill bits), but drilling progress is slowed.   Boulders up to one-third the shaft diameter can usually be removed by special tools—"boulder bits" or buckets, or "grabs" of various designs.   Larger boulders can sometimes be broken up by "gads" or other tools, but more commonly require blasting or manual excavation using a jackhammer—a very expensive procedure.   Also, entering a hole drilled through a bouldery or cobbly formation may be dangerous, necessitating the extra expense of temporary or permanent casing.   It should be remembered, however, that boulders also give much trouble in the installation of any other type of deep foundation, and that when a drilled pier hole has once succeeded in bypassing them to an assigned bearing formation, the foundation inspector may be sure that suitable bearing has been reached.

Water Entering the Hole

Water may enter the pier hole as it is being drilled or after completion, either from pervious (water-bearing) layers in the soil or from cracks, fissures, or joints in rock. The water must be coped with in some manner—either by sealing it off or by pumping, so that an essentially dry hole can be prepared for concreting; or alternatively, by employing special below-water deposition of concrete. Either method is expensive, and circumstances which make coping with a substantial volume or head of water necessary may add very substantially to the cost of construction. Moreover, unless the water is controlled, seepage forces will tend to cause collapse, or may produce extensive soil "piping" and subsequent caving of sidewalls. The resultant "ground loss" may cause subsidence of the adjacent surface, as well as overrun of concrete quantities and possible uncertain lateral support of pier shafts.

Lack of Suitable Bearing Formations

Sometimes local geology is such that a stratum or strata suitable for end-bearing or sidewall shear, or for belling where bells would be required, are not present, and consequently another type of foundation has to be selected. An example is found in the New Jersey Meadows, where typically 6 to 11 ft of soft organic marsh deposit overlie a varved glacial lake-bed deposit which may be up to 150 ft (46 m) or more in thickness. The top few feet of the varved deposit constitute a "crust" of stiff, preconsolidated soil which is relatively strong and incompressible; the varved deposit below the crust is soft to very soft. The crust would furnish moderate bearing values for underreamed piers, but underreams could not be formed in the overlying marsh deposit. They could be formed in the crust itself, but their bottoms would then be too close to the underlying soft soil. Consequently, drilled piers are not used in this region, and other foundation systems have been adapted to the local conditions.

Variable Subsurface Conditions

Under some geologic circumstances—especially in formations such as glacial outwash deposits, braided stream deposits, and residual soil formations (especially those derived from solution-susceptible rock)—bearing conditions and levels are so variable within a distance of a few feet that interpolation between test borings is often pure speculation. These are usually deposits containing some conditions that are unfavorable to the construction of drilled piers. Even when drilled piers have been installed successfully at nearby locations, the possible varia-tion in bearing level in such formations is so great—and so likely—that drilled pier foundations may be a risky choice, particularly with respect to budgeting and construction costs.

Piers which depend on rock for support have to penetrate far enough into sound rock and—where irregular rock quality is suspected—must be proof-tested, at additional cost, to prove the presence of acceptable rock in the zones which will be highly stressed.   This requirement often means penetration of hard rock zones to bypass excessively seamed or fractured rock or solution voids.   Cavernous, fissured, or pinnacled rock can rarely be drilled without special tools.   Sometimes a core barrel can be used with the auger machine, or a Calyx drill (see Chap. 4) can be used, but usually the rock will have to be broken up by blasting or by jackhammer work performed in the pier hole.   Such "dental work," as it is sometimes called, is very expensive, 1971 costs of $200 to $400 per cu yd of rock removed being common.

Another difficulty that applies to drilling operations for rock-bearing piers is that it is often difficult to seal a casing into the top of rock to cut off water entry from the weathered zone or from a gravelly or bouldery layer immediately overlying bedrock, and it may not be feasible to cut off water entry from fissures or caverns encountered at lower levels.   As always, if water cannot be removed, concrete must be placed by pumping or by tremie, an additional expense, and the advantage of direct inspection of the bearing material may be lost.   Conditions like these are particularly prevalent in limestone or dolomite bedrock, as well as in other rock types susceptible to solution activity.

The unfavorable conditions described above apply also to other types of deep foundation; and because the bearing surfaces usually can be examined in a pier hole, this type of foundation is often the choice despite the high (and unpredictable) cost of "dental work."   (However, averaged conditions over such a site are often suitable for support of a shallow mat-type foundation, and careful cost analysis will sometimes lead to the choice of this variety instead of any type of deep foundation.)

Miscellaneous, Trash, or Stony Fill

Controlled fills of either cohesive soils or compacted sand can usually be drilled readily and will stand open nicely.   However, uncontrolled fills and compacted fills containing rubble, boulders, or large cobbles should always be considered potentially troublesome.   An uncontrolled fill may contain anything—for example, an innerspring mattress or an old automobile body—completely surrounded by easily drillable soil. Although a contractor may be quite confident that he can get piers down economically—and be quite right for nine out of ten of them—the few that do turn out to be troublesome may cost more and take more time than all the rest together, and the job will "turn sour."

An expedient sometimes found economical is to excavate the fill at each pier location, set casing on the natural subsoil, backfill around the casing, and then drill.

Prolonged Bad Weather

Because pier-drilling and concreting operations are more difficult in bad weather than are most pile-driving operations (several piles can be driven from one rig station), seasons of heavy rain or snow should be taken into account in selecting the type of foundation system to use.

Questionable Equipment or Contractor Competence

In instances where there is question about the ability of available drilled pier contractors to do the required work efficiently and a tight construction schedule has been planned, with substantial extra costs accruing from construction delays, it is sometimes prudent to use another type of foundation for which scheduling can be more certain.

2-4   BORDERLINE CONDITIONS

For sites where both favorable and unfavorable conditions exist, so that it is difficult to judge which will prevail, a large-scale drilling test can be set up to test the effect of the unfavorable factors. If this cannot be done, another foundation type should be selected. When substantial doubt of feasibility exists and a contractor insists that he can do the job, he may be invited to prove it—at his expense. Large-diameter test borings for this purpose should be located where the earlier test borings have shown the most unfavorable ground conditions—but it should be kept in mind that still more unfavorable conditions may exist elsewhere on the site.

A tendency for underreams to cave is a condition that may not become evident until actual construction has started. This may be rated as a borderline condition; caving may occur in only a few holes, with all the others being underreamed without trouble. In such cases it is desirable to allow for design alternatives which permit a change either to a deeper straight shaft, supported by sidewall shear, or to a straight shaft of larger diameter to gain sufficient end-bearing. A shaft of smaller diameter, retained by a suitable liner, may be used above the base "footing" to reduce concrete quantities. The change need be made only for those piers that give trouble during construction. It is quite feasible—and sound engineering—to mix types of deep foundation within a single structure, as long as the designs are such as to keep differential settlements between piers within allowable structural tolerances (see Art. 3-3).

Another condition that may reach the "borderline" case during construction is difficulty in controlling water entering the pier hole. Sometimes most of the pier holes on a job will give no trouble in this respect, but a few holes will not respond to routine means for sealing off water or keeping it pumped down. There are many expedients that can be tried in an effort to control water in such cases, but it is usually better to go at once to a positive, though more expensive, procedure. Examples would be to complete the drilling in a hole filled with water (to assure outflow rather than inflow) and then deposit the concrete by tremie or "elephant's trunk," using extra-strength concrete, or to bring in a more powerful drilling machine to screw casing down to a seal through cobbles or weathered rock. In general, borderline conditions require positive solutions, not marginal or experimental remedies.

2-5 SPECIAL CONSIDERATIONS

Completion Time

Experience has shown that when drilled piers are installed successfully but with difficulty under borderline conditions, the time required to do the work may be so long that added indirect expenses—e.g., inspection costs, interest charges, loss of revenue—far exceed the difference in cost between the pier foundations and a more readily constructed type of foundation.

Cap Costs

A cost item that must be considered when making comparisons between the cost of drilled pier foundations and other types of foundation is that of the pile caps for pile foundations. Drilled piers are usually (though not always) designed as a single unit, with dowel rods in the top for connection to column bases. Piles, on the other hand, are nearly always in groups; the transfer of load from the structure is through a reinforced concrete pile cap, which is an item of considerable expense.

Ground Heave vs. Loss of Ground

Pier-drilling operations do not result in ground heave, whereas the driving of displacement piles may, under certain circumstances, produce very troublesome heave and horizontal displacement of ground and/or adjacent piles. Pier-drilling operations, however, can produce loss of ground, but this is usually controllable. In general, there appears to be

Fig. 2-1 Loss of ground around a pier hole. Here spoil was blown out of the hole continuously by compressed air, and inflow of water and silt below the casing caused surface subsidence.

less danger of trouble from loss of ground from pier drilling than of heave produced by driving closely spaced displacement piles [2.1]. The comparison is, of course, relevant only when both systems could be used on the site.

Danger of Quantity Overrun

In some areas, overruns in depth, caused by failure to stop the drilling when suitable bearing depth was reached, have caused major increases in foundation costs and have led to reluctance on the part of some architects and engineers to specify this type of foundation. The remedy lies in adequate and competent foundation investigations, plus careful inspection. Generally, the potential saving can far exceed the cost of making sure that the work is done properly.

2-6 SELECTION OF PIER TYPE

The performance, and particularly the economy, of a drilled pier foundation system is often strongly dependent on the compatibility of the type of pier selected to conditions imposed by the project environment. In the authors' opinion, the two most significant of the several factors controlling the selection of a pier type are the prevailing ground conditions and the magnitude and orientation of the imposed pier loads.

Underreamed (Belled) Piers

Underreamed piers are widely used where the imposed pier loads are moderate to heavy and where a high-capacity end-bearing stratum can be reached at a reasonable depth without excessively hard drilling. The underreamed pier also provides efficient resistance to uplift loads and can—for short, rigid piers—substantially increase the resistance to lateral loads. Costs can often be minimized by limiting the concrete shaft diameter to only that necessary to carry the required loads, although the shaft hole must be large enough to permit cleanout and inspection—at least 24, preferably 30 in. (61 and 76 cm) in diameter.

The difficulty of construction and the associated costs of underreamed pier construction increase markedly where the materials immediately overlying a suitable bearing stratum are subject to collapse during underreaming. This condition has been often encountered in the authors' practice, where a relatively sound rock is encountered beneath an alluvial or glaciofluvial overburden, and also where residual silts and clays overlie solution-prone rocks. The presence of a high groundwater level will also inhibit underreaming by inducing "blow-ins" or progressive collapse if a "water seal" cannot be obtained above the underreaming level.

As discussed in Art. 4-2, "mud slurry drilling" or other special drilling techniques can be used to allow underreamed pier construction in unstable soils—at an added cost. Underreaming below casing carried through unstable materials and seated on rock has also been performed by heavy-duty, high-torque drills or by blasting and/or using pneumatic tools to excavate the bell. However, in relatively competent rocks, these techniques are rarely competitive with the use of straight piers supported by sidewall shear as well as end-bearing.

Straight-Shaft Piers

Drilled piers with a uniform section are applicable for relatively light pier loads, and even for heavily loaded piers which carry the greatest portion of the imposed load by shearing resistance between the shaft and the surrounding soil or rock, i.e., shaft resistance or "sidewall shear." Sidewall shear will also provide resistance to uplift, especially if the walls are rough or have been artificially grooved or roughened.

The "straight-shaft" pier is also often used as an end-bearing unit where subsurface conditions are not favorable to underreaming. In these circumstances, the hole diameter is made large enough to develop the required end-bearing area, and the shaft above the bearing section can be limited in diameter by the use of a Sonotube or light metal casing. Alternatively, a straight-shaft pier can be socketed into the bearing stratum (with a socket diameter substantially the same as that of the shaft) to develop both base and shaft resistance. Where the socket is in rock, the added cost of rock drilling and increased shaft length must be weighed against the added cost of soil drilling and the additional concrete required by a large-diameter straight-shaft pier bearing at a higher elevation.

The pier socketed into a relatively competent rock has the disadvantage of requiring special percussion, or heavy-duty auger drilling equipment capable of delivering a torque of at least 80,000 to 100,000 lb-ft (109,000 to 136,000 newton-m), to drill the rock socket. Further, machines of this type are sometimes not locally available through more than one contractor, and "machine sockets" may inadvertently become a costly proprietary construction. The cost of manual socket excavation using blasting and pneumatic tools also entails substantial cost premiums. Unless the quality of the bearing stratum significantly increases with depth and a relatively high sidewall shear value can be justified, either underreaming in rock or a larger-diameter straght-shaft pier may prove to be economically competitive to rock-socket construction. On several projects in the authors' practice both socketed and underreamed piers have been used interchangeably as dictated by rock conditions. On these projects the socketed piers were used only in those locations where underreaming proved to be technically or economically unfeasible.

REFERENCES

2.1 Casagrande, L., Subsoils and Foundation Design in Richmond, Virginia, *Proc. ASCE*, vol. 92, SM 3, pp. 109–126, September, 1966.

CHAPTER THREE

Design Considerations

3-1 SURVEY OF DESIGN PRACTICE

Like many other areas of engineering design involving substructure-subgrade interaction, despite the rapidly advancing state of knowledge, the state of the art of deep foundation design must presently be considered to be at best a semiempirical technique. Further, a review of practice in America and Great Britain indicates that although design procedures employing theoretical considerations to varying degrees are widespread, a substantial amount of pier design remains entirely empirical, with no basis in theory.

3-1.1 Semiempirical Design

The most commonly used semiempirical approach to the design of axially loaded piers is based on prediction of the ultimate load capacity of a single pier, from which the working (design) loading is determined by application of a suitable factor of safety. A further reduction in design capacity for piers in groups is often made to allow for the "efficiency" of the group. This design method will be considered for compressive and tensile loads in Arts. 3-3 and 3-6 as "Limit (Ultimate Load) Design Analysis."

Although not currently in wide use, more theoretically rigorous design techniques for axially loaded piers have been derived from consideration of the compatibility of pier load and displacement and are considered in Art. 3-3. Contrary to the general design practice for axial loads, current design methods applicable to laterally loaded piers are frequently based on displacement compatibility techniques. These techniques, as well as the limit design approach to lateral load capacity, are summarized in Art. 3-5.

3-1.2 Empirical Design

Many piers, particularly where rock bearing is used, have been designed using strictly empirical considerations which are derived from regional experience or from the results of prototype load tests. Where subsurface conditions are well established and are relatively uniform, and the performance of past constructions well documented, the "design by experience" or by "precedent" approach is usually found to be satisfactory—particularly when the experience has been obtained from systematic observations of the performance of piers with full knowledge of the controlling load and subsurface conditions. Ideally, regional performance data are accumulated using a long-term design, construct, and observe program. By this means successively higher design capacities can be employed and optimum design criteria eventually established by performance correlations. The obvious limitations to "design by experience" relate to the inability to establish adequate performance correlations in areas where subsurface conditions are variable and where the loads to be supported have little or no precedent.

In some instances, regional experience may be reflected in local building codes in the form of "presumptive bearing values" which are prescribed for various material types. However, the use of bearing values derived from building codes without knowledge of the derivation and limitations thereof is, at best, ill-advised and has too frequently led to unsatisfactory performance or uneconomical designs.

Typical bearing values, primarily derived for piers drilled into rock, are included in Appendix E as a sampling of regional practice in the United States.

3-1.3 Load Testing

The results of prototype load tests are sometimes applied to pier design by assuming a direct correlation with the test pier. Load-test results are also used—perhaps more commonly—to provide shearing resistance and bearing parameters applicable to analytic design methods. Several load-testing techniques have been developed, ranging from axial and lateral load tests on full-scale piers to tests on small-diameter bearing plates placed in the bottoms of excavated shafts. As outlined in Art. 3-3, a variety of interpretative techniques are also currently used which involve both conventional and specially instrumented load tests.

The cost of a load-testing program is relatively high, and the number of tests that can be performed is usually limited. For these reasons, load testing is generally applicable only to design situations where subsurface conditions are relatively uniform and where a limited amount of testing can be considered to be reasonably representative of the site. However, specially instrumented load tests, together with complete documentation of the subsoils and of the installation techniques, are sorely needed to advance the basic understanding of deep foundations.

3-2 CONCEPTS OF AXIAL LOAD TRANSFER

The transfer of axial load from a loaded pier to the supporting material is dependent on many interrelated factors such as subsurface conditions, the geometric and structural features of the pier, the method of construction, and the elapsed time after construction. Some of these interdependent factors which have been the topic of study are:

1. The shearing (shaft) resistance and bearing (base) resistance as functions of shaft and base displacement
2. The relationship between the ultimate shaft resistance and the initial undrained shear strength of the materials in contact with the shaft
3. The stiffness (or compressibility) of the pier as compared to the supporting material
4. Pier dimensions (length/width) and configuration (belled or straight-shaft)—singly and in groups
5. The migration of water from fresh concrete to fine-grained soils in contact with the shaft

Other factors influencing the long-term performance of drilled piers— such as drilling disturbance, time of exposure before concreting, sidewall smear, imperfect bottom cleanout, and seasonal changes within the active zone of the soil mantle,—cannot be quantitatively evaluated and can presently be reflected in design only by the application of "good engineering judgment." Nothwithstanding the many indeterminate factors which can influence pier performance, consideration of how loads are transferred from pier to ground can serve as a basis for a more knowledgeable evaluation of design parameters—regardless of the design method used.

Despite the wishes of those designers who insist on a "settlement-free" foundation, some downward movement of a pier is required to mobilize either shearing resistance around the shaft or bearing resistance at the base of the shaft. During the initial stages of loading, the downward movement of the shaft and of the materials in contact with the shaft are compatible. With continuing shaft movement, the ultimate shearing resistance of the soil is eventually fully mobilized, and further movement causes slippage of the pier relative to the soil along or in the proximity of the shaft surface.

Usually, the ultimate shearing resistance of a pier of normal stiffness and length will be reached first near the top, where shaft displacement is the greatest and the confinement the least, and will subsequently extend downward. Finally, as the load increases, the movement of the pier relative to the soil is sufficient to fully mobilize the ultimate shearing resistance of the soil all along the shaft. In some materials, the shaft resistance may reach a peak value and subsequently drop off to a relatively constant value with increasing shaft movement.

3-2.1 Division of Load Between Shaft and Base

Because full sidewall shearing (shaft) resistance is developed at much smaller axial movements than is end-bearing, appreciable shaft resistance may be developed before significant load can be transferred to the base, particularly in the case of long, slender piers. Thus, when shaft resistance has been fully mobilized, all additional load is transferred to end-bearing until the material supporting the base is overstressed and failure occurs.

Observations made by Whitaker and Cooke [3.1] during pier load tests conducted in London clay indicate that full shaft resistance is developed in stiff fissured clays after movements of the order of ¼ in. (about 6 mm). From pier tests in stiff Beaumont clay, Reese and O'Neill [3.2] found that a similar displacement, about 0.2 in. (5 mm), was required for mobilization of full shaft resistance. Vesic [3.3] reports that displacements of as much as 0.4 in. (10 mm) are needed to mobilize full shaft resistance in sands and that such displacements appear to be independent of shaft dimensions for both sand and clay.

The amount of displacement necessary to mobilize full base resistance will be dependent on the type and confinement of the bearing materials, but for a given bearing condition it appears to be a function of the base diameter. Instrumented load tests have shown that the amount of base movement required to develop ultimate end-bearing capacity ranges from as much as 25 percent of the base diameter for cohesive soils to 8 to 10 percent for cohesionless materials. Load tests in stiff clays (by far the majority of all such tests made) indicate base displacements at failure to be of the order of 10 percent of the base diameter.

The division of load between shaft and base for a straight-shaft pier drilled in stiff clay is shown clearly at various levels of applied load in Fig. 3-1. This 36-in.-diameter (91 cm), 25-ft-long (7.6 m) pier was instrumented by telltales (see Art. 3-3) to show the vertical movement and the load carried at various depths along the shaft. Observations were made for a series of applied loads. These show the first two increments to be carried almost entirely by shaft resistance (sidewall shear). For loads greater than about 80 tons (73,000 kgf), the ultimate shaft resistance is developed along the entire length of the shaft, and additional applied load is taken entirely by base resistance as demonstrated by the parallel orientation of the final two load-distribution curves.

The load-distribution curves and the corresponding shaft deformation can also be analyzed to determine the relationship between shearing resistance at a point or interval along the shaft, and shaft movement, i.e., the load-transfer relationship. Pile and pier load-test analyses of this type have been made by Seed and Reese [3.4], D'Appolonia and

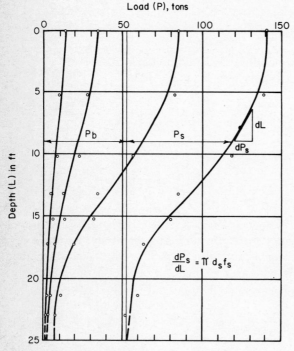

Fig. 3-1 Load distribution in a 25-ft-long (7.6 m) straight-shaft pier drilled into stiff clay. Note the reduced rate of load transfer (dP_s/dL) near the top and base of the pier. [*After Reese and O'Neill* (*1969*).]

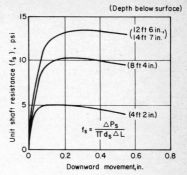

Fig. 3-2 Load transfer to shaft segment (ΔL) vs. downward movement of segment; unit shearing resistance (f_s) calculated from load-distribution curve (Fig. 3-1). [*After Reese and O'Neill (1969)*.]

Romualdi [3.5], DuBose [3.6], and others. Typical results of a load-transfer analysis of a 36-in.-diameter (91 cm) instrumented pier are shown in Fig. 3-2, demonstrating the influence of depth and deformation on shearing resistance. Both Figs. 3-1 and 3-2 indicate a reduction in the rate of load transfer in the lower part of the shaft, particularly just above the base.

Several investigators report that a reduced rate of load transfer near the base is also typical of piles and piers in sand, and attribute the phenomenon to reduced lateral stresses in the vicinity of the base, due to soil arching, or to radial stresses induced by the base load. Hence, the assumption often made in the limit design method—that the shaft and base resistance of granular soils continues to increase with an increase in overburden pressure—is not substantiated by field measurements. The additional assumption commonly made—that the shaft resistance of cohesive soils can be expressed as a predictable percentage of the undrained shear strength, independent of shaft movement—has also been repeatedly demonstrated by instrumented field load tests to be an oversimplification.

3-2.2 Influence of Pier Dimensions and Stiffness

Load transfer from a pier to the surrounding materials, and the corresponding pier settlement, are governed to a large degree by the length-to-diameter ratio of the shaft, the ratio of the shaft diameter to the base diameter, the relative stiffness of the pier as compared to the stiffness of the surrounding material, and the relative stiffness of the materials above and below the base. The influence of these parameters has been investigated by Mattes and Poulos [3.7, 3.8] using mathematical models of single "compressible" and "incompressible" piers or piles embedded within an elastic isotropic medium of both semi-infinite and finite extent.

As indicated by Fig. 3-3, an elastic solution for a rigid pier which con-

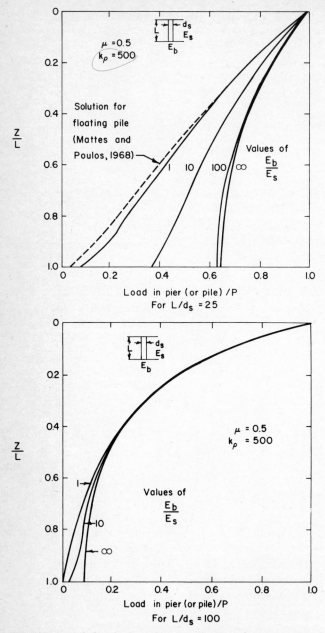

Fig. 3-3 Load distribution along a pier (or pile) in a semi-infinite elastic mass as a function of the ratio of the modulus of the materials around the shaft to that of the materials below the base. [*After Poulos and Mattes (1969).*]

siders a difference in the relative compressibility of the materials around the shaft and below the base of a pier demonstrates that both an increase in the ratio of the modulus of the base-supporting material (E_b) to that of the shaft-supporting material (E_s) and a decrease in the shaft length-diameter ratio (L/d_s) increase the proportion of load transferred to the base. Figure 3-4 shows, for a pier in an elastic medium, that the proportion of load transferred to the base increases with an increase in the rigidity of the pier shaft and is dependent on L/d_s.

From theoretical investigations of piers with enlarged bases in an ideal elastic medium, Poulos and Davis [3.9] found that the base diameter (d_b) had a significant influence on pier behavior only when L/d_s is less than about 25. If elastic behavior of the bearing materials is assumed, it can be concluded that underreamed piers would not offer any appreciable load-bearing advantage over straight-shaft piers of equal shaft diameter and length where the L/d_s of the pier is greater than about 25, *unless* the end-bearing stratum is significantly stiffer than the overlying materials. Mattes' and Poulos' work also demonstrates, as shown by Fig. 3-5, that as k_p decreases, the load transferred to the base decreases markedly, becoming more pronounced at greater d_b/d_s values.

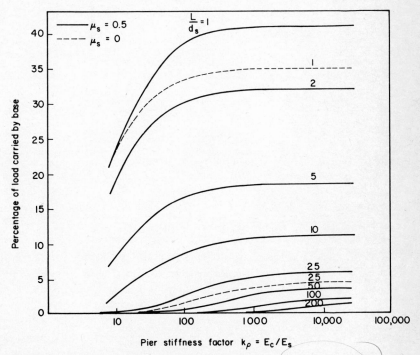

Fig. 3-4 Pier load carried by base as a function of shaft rigidity (k_p) and slenderness (L/d). [*After Mattes and Poulos (1969).*]

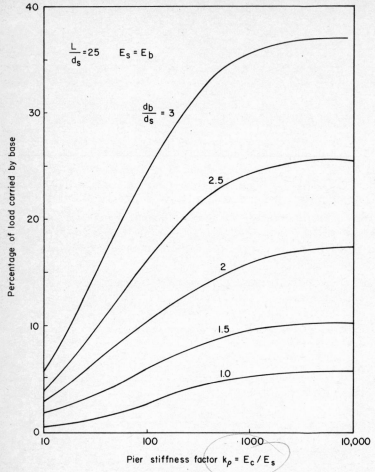

Fig. 3-5 Load carried by enlarged base piers in semi-infinite elastic mass.
[*After Mattes and Poulos (1969).*]

3-3 DESIGN FOR COMPRESSIVE LOAD

As cited in Art. 3-1, there are at least four design approaches currently
being applied, singly or in combination, to the design of drilled piers
for compressive load. Of these methods, the widely used limit design
technique is given primary attention and is followed by a brief consideration of a design method using displacement compatibility concepts.
Load-test interpretation techniques are also reviewed, and examples of
pier design parameters, developed primarily through regional experience,
are presented in Appendix E.

3-3.1 Limit Design (Ultimate Load) Analysis

To determine a "safe" working load by limit analysis, the ultimate load which can be sustained by a single pier (P) can be calculated as the sum of the ultimate shaft resistance (P_s) and the ultimate base resistance (P_b), each reduced by an appropriate factor of safety $(F_s$ and $F_b)$. This relationship is described by the following expression, where p_b and f_s are unit base and shaft resistance applied over the engaged shaft area (a_s) and base area (a_b).

$$P = \frac{P_b}{F_b} + \frac{P_s}{F_s} = \frac{a_b p_b}{F_b} + \frac{a_s f_s}{F_s} \tag{3-1}$$

In American practice, the shaft resistance is often entirely ignored for underreamed piers, particularly if the piers are designed to bear on a stratum which is relatively rigid when compared to the overlying materials, whereas usual British practice is to design belled piers for both shaft and base support. Conversely, piers of uniform section drilled into clays and shales have been designed exclusively for shaft resistance, with the end-bearing component considered as an added factor of safety. The validity of these assumptions has been shown to depend upon the pier dimensions and properties $(L, d_s, d_b,$ and $E_c)$, as well as on the unit loads and the properties of the supporting materials $(P_b, f_s, E_s,$ and $E_b)$.

Ultimate shaft resistance is calculated as the sum of the ultimate shearing resistances imposed by the various strata (or increments thereof) which are in contact with the pier. The ultimate shaft resistance of a given shaft segment (ΔL) is commonly expressed for cohesive materials in terms of the undrained shear strength (s_u) and a reduction factor (α). For cohesionless materials, ultimate shaft resistance is expressed in terms of the effective vertical pressure (σ'_v), the coefficient of lateral earth pressure (k_s), and the tangent of the angle of frictional resistance between the shaft and the surrounding materials $(\tan \delta)$. These relationships are commonly expressed as

Cohesive: $\qquad P_s = \Sigma \pi d_s \, \Delta L \alpha s_u \tag{3-2}$

Cohesionless: $\qquad P_s = \Sigma \pi d_s \, \Delta L k_s \sigma'_v \tan \delta \tag{3-3}$

It is significant that Johannessen and Bjerrum [3.10], Chandler [3.11], and others have suggested that a form of Eq. (3-3), employing effective stress parameters, may also be appropriate for prediction of the shaft resistance of piles (or piers) in cohesive soils.

A great many solutions, usually based on plastic equilibrium concepts, have been developed to express the "bearing capacity" of foundations.

Solutions most frequently used for deep foundations are those of Meyerhof [3.12], Terzaghi [3.13], and Berezantsev [3.14]. These solutions involve dimensionless bearing capacity factors (N_c, N_q, N_γ, and N_q'), unit soil weight (γ), pier depth (L), and in some areas of practice a reduction factor (ω) to be applied to the shear strength of cohesive bearing materials. The most widely used solutions are expressed for cohesive and cohesionless materials as

Cohesive: $\qquad\qquad P_b = a_b(N_c \omega s_u + \gamma L)$ $\qquad\qquad\qquad\qquad$ (3-4)

Cohesionless: $\qquad\quad P_b = a_b(0.4\gamma_1 d_b N_\gamma + \gamma L N_q)$ $\qquad\qquad\quad$ (3-5)

Cohesionless: $\qquad\quad P_b = a_b(\sigma_v' N_q')$ $\qquad\qquad\qquad\qquad\qquad$ (3-5a)

The second term in Eq. (3-4) is usually ignored in bearing capacity calculations, considering the overburden pressure (γL) to be offset by the weight of the pier, which is not included in the imposed load calculation.

The factor γ_1 in Eq. (3-5), as derived by Terzaghi, is quite sensitive to the volumetric compressibility of the soil, and its evaluation is uncertain. In practice this has led to the assumption that $\gamma_1 = \gamma$. Except for relatively short piers, the first term in Eq. (3-5) will be comparatively small and is also often ignored in base resistance calculations. It also is noted that σ_v' in Eq. (3-5a) represents the effective vertical pressure at the level of the pier base, and that σ_v' is not necessarily equal to the overburden pressure should subsoil arching be mobilized around the pier base.

(a) Shear Strength Parameters for Cohesive Soils The undrained shear strength of cohesive materials is most often determined in the laboratory from unconsolidated-undrained triaxial and uniaxial compression tests on representative "undisturbed" samples. Alternatively, the undrained shear strength of cohesive soils having a soft-to-firm consistency can be evaluated in situ from field vane shear tests. Other less direct or more approximate methods which involve in-situ shear strength evaluation include the pressuremeter, cone penetration resistance, and standard penetration resistance testing. DeMello [3.15] presents a comprehensive review of testing techniques and interpretations applicable to the evaluation of the shear strength of cohesive soils.

Only sparse data are available concerning the ratio of the undisturbed undrained shear strength of a cohesive soil, as measured by laboratory or field shear tests, to the ultimate shearing resistance (adhesion) which is mobilized along a pier shaft. Average values of this empirical ratio, termed the shear strength reduction factor (α), as derived from analysis of prototype field load tests in clays, shales, and tills, are listed in Table 3-1. On the basis of analyses of load tests on piers and piles in cohesive

TABLE 3-1 Shear Strength Reduction Factors (α) for Drilled Piers

Material	Material properties			α	Reference
	w_n, %	l_p, %	s_u, tsf or kgf/cm^2		
Stiff clay..........	23	35–55	1.2	0.44	Whitaker and Cooke [3.1]
Stiff clay..........	25	20–60	1.2	0.62	Reese and O'Neill [3.2]
Massive shale.....	15	7–16	5.0	0.64*	Matich and Kozicki [3.24]
Glacial till........	12	2–16	2.5	0.64*†	Matich and Kozicki [3.24]
Stiff clay..........	...	.....	1.1	0.52	Woodward et al. [3.17]
			0.9	0.49	
Stiff clay..........	19	36–46	1.4	0.30	Mohan and Jain [3.25]

* Failure was not reached.

† Sandy gravel with cobbles and approximately 50 percent silty clay, $N \geq 45$ blows per ft.

soils [3.16, 3.17], it is generally accepted that α increases with a decrease in undrained shear strength, as shown by Fig. 3-27. For drilled piers, α is also to a great extent influenced by construction effects and the moisture sensitivity of the supporting materials. The reduction of the undrained shear strength of a stiff clay due to migration of water from fresh concrete has been investigated by Chuang [3.18], who proposes a method to directly evaluate α from laboratory tests. Reductions in α due to construction procedures and the exposure duration of open shafts are also discussed for stiff fissured London clays by Whitaker and Cooke [3.1] and Burland et al. [3.19].

(b) Shear Strength Parameters for Cohesionless Soils Triaxial and direct shear tests, generally on reconstituted samples, have been used to estimate the ultimate shearing resistance of rough concrete surfaces embedded in granular soils as a function of the friction angle of the soil (ϕ), assuming $\phi \simeq \delta$ in accordance with Eq. (3-3). The ratio δ/ϕ for metal liners of the type which would be left in place can be conservatively assumed to range from 0.7 (smooth) to 0.85 (corrugated).

The work of Kerisel [3.20] and Vesic [3.21] with both piles and piers indicates that the shearing resistance along a shaft in granular soil is a linear function of overburden pressure only to a limited depth, becoming approximately constant below a depth of about $10d_s$ in very loose sand to $20d_s$ in very dense sand. Vesic's work also indicates that k_s decreases slightly with an increase in shaft size for piers with a length less than about $20d_s$, but that k_s is dependent on relative density (D_r) regardless of the pier length. Although few data are available concerning the appropriate range of k_s for drilled piers, it is probable that k_s

does not significantly exceed the coefficient of earth pressure at rest (k_0), expressed as

$$k_s \simeq k_0 = 1 - \sin \phi \qquad (3\text{-}6)$$

and is not less than the coefficient of active earth pressure (k_a).

More commonly, the unit shaft resistance (f_s) in granular soils is estimated by empirical correlations with standard penetration resistance (N) in American practice, and by the frictional component (f_c) of static cone penetration in European practice. Meyerhof [3.22] has expressed shaft resistance in sands as a function of the standard penetration and cone resistances as

$$f_s = 0.01N \leq 0.5 \text{ tsf (kgf/cm}^2) \qquad (3\text{-}7)$$

$$f_s = f_c \leq 0.5 \text{ tsf} \qquad (3\text{-}8)$$

where N is in blows per foot.

On the basis of empirical correlations with model and full-scale load tests, Vesic [3.23] has also derived the following expression for the ultimate shaft resistance (in tsf) of saturated or dry, normally consolidated sand as a function of relatively density only:

$$f_s = 0.025(10)^{1.5D_r^4} \qquad (3\text{-}9)$$

(c) Bearing Capacity Factors The N_c factor for piers or piles in cohesive materials bearing at depths greater than four base diameters has long been accepted, on the basis of both theoretical and experimental evidence, to be approximately equal to nine. For lesser depths, N_c is shown by Skempton [3.26] to be dependent on the ratio of depth to base diameter. In Great Britain, a reduction factor (ω) has been applied to the undrained shear strength of heavily overconsolidated and fissured clays to allow for test sample scale and disturbance effects. For example, $\omega = 0.75$ is used in fissured London clays for piers with base diameters ≥ 3 ft (92 cm). It is likely that ω can be taken as unity when good "undisturbed" samples of nonfissured, cohesive materials are considered.

Bearing capacity factors according to Terzaghi, Berezantsev, and Meyerhof are shown in Fig. 3-6 as a function of the effective friction angle (ϕ). Except at relatively shallow depths, Eqs. (3-5) and (3-5a) have often been reported to considerably overestimate end-bearing. However, the common and incorrect assumption that $\sigma_v' = \sigma_0$ is probably the principal cause of the discrepancy between observed and predicted values.

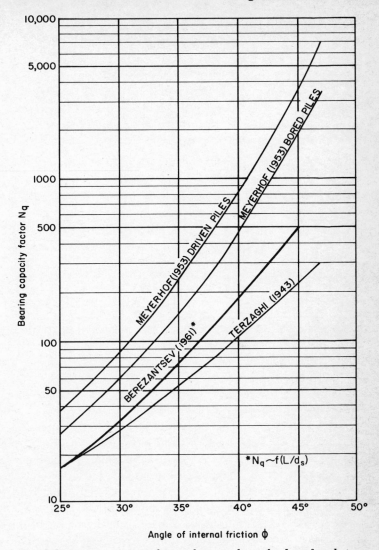

Fig. 3-6 Bearing capacity factors for granular soils, deep foundations, circular bearing area.

The most rational approach for estimating the end-bearing of drilled piers founded on granular subsoils presently appears to be that expressed by Eq. (3-5a), using N_q' after Berezantsev (most conservative) or Meyerhof (least conservative). The σ_v' term of the equation should be calculated as the effective overburden pressure at a depth of 15 to 20

TABLE 3-2

Soil type	q_c/N
Silt and sandy silt......................	2.0
Sand and slightly silty sand............	3.5
Gravelly sand.........................	5.0
Sandy gravel and gravel................	6.0

pier diameters, depending on the relative density of the materials in contact with the shaft. For piers shorter than about 15 diameters, σ'_v would be equal to σ_0 at the pier base, assuming that soil arching would not be significant above this depth.

Empirical correlations for p_b in tsf are proposed for sands by Meyerhof [3.22] in terms of the standard penetration resistance (N) and static cone point resistance (q_c) as

$$p_b = 4N \tag{3-10}$$

$$p_b = q_c \tag{3-11}$$

These relationships assume that $L/d_s > 10$ and that the subsoils are saturated sands. As the p_b vs. N relationship of Eq. (3-10) is based on empirical correlations between N and q_c, it is pertinent to note that Schmertmann [3.27] considers q_c/N to be a function of the texture of the subsoils, in accordance with Table 3-2.

(d) Settlement Analysis Before any analysis of pier settlement can be attempted in conjunction with limit design techniques, a division of the pier load between shaft and base resistance must be made. As discussed in Art. 3-1, this division at working loads is not simply the ratio of the ultimate shaft to base resistance but must consider, in some fashion, the displacement compatibility of shaft and base resistance.

The settlement of the top of a single drilled pier (Δ_p) can be approximately described, in terms of the elastic shortening of the shaft (Δ_s) and the settlement of the base due to base load (Δ_{bb}) and shaft load (Δ_{bs}), as a function of the base and shaft resistances $(P_b$ and $P_s)$ and the material moduli (E_b, E_s, E_c). The various components of settlement are expressed by the following equations:

$$\Delta_p = \Delta_{bb} + \Delta_{bs} + \Delta_s \tag{3-12}$$

$$\Delta_{bb} = \frac{p_b d_b I_{bb}}{E_b} \tag{3-13}$$

$$\Delta_s = \frac{(P_b + \lambda P_s)L}{a_s E_c} \tag{3-14}$$

$$\Delta_{bs} = \sum \frac{\Delta P_s I_{bs}}{E_b L} \tag{3-15}$$

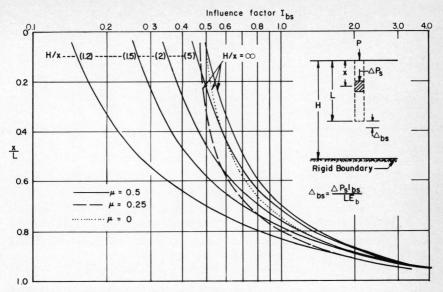

Fig. 3-7 Base displacement for shaft load segments.

The base deflection influence factors I_{bs} and I_{bb}, as derived from the integration of the Mindlin solution for a point load in the interior of an isotropic elastic half-space, are shown in Figs. 3-7 and 3-8, respectively. The influence of shaft resistance on base deflection can be expressed in accordance with Eq. (3-15) by dividing the shaft load into a series of point loads (ΔP_s) acting at the center of each of the associated pier segments. More rigorous solutions, simulating shaft load segments by incremental line or ring loads, are available as demonstrated by Poulos and Davis [3.9]. However, considering the broad approximations in-

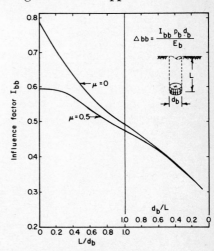

Fig. 3-8 Displacement of rigid base with uniform bearing pressure.

volved in Eqs. (3-13) to (3-15), and that the base deflection due to shaft loading Δ_{bs} probably accounts for less than 10 percent of the pier settlement, the added computational effort is not justified.

The parameter λ of Eq. (3-14) expresses the ratio of average to total shaft load, is a function of the shape of the load-distribution curve as shown by Fig. 3-14, and is influenced by long-term creep or consolidation effects. Assuming limiting parabolic curve forms such as are often observed in instrumented pier and pile load tests, λ would usually be expected to be between $\frac{1}{3}$ and $\frac{2}{3}$. A value of 0.6 has been suggested by Vesic [3.3] for single piles in sand. Integration of pier load distribution curves published by Reese and O'Neill [3.2] indicates λ to be approximately 0.5 for piers drilled into stiff Beaumont clay.

Although theoretical considerations indicate that the greater part of single pier settlement occurs as an immediate distortion, combined shear and volumetric (consolidation) settlement can be estimated from Eqs. (3-12) to (3-15) by deriving E_b and E_s in terms of effective stress from laboratory shear tests, ideally following stress paths simulating field loading conditions. Immediate settlement can be estimated from undrained shear test results for representative "undisturbed" samples of the bearing materials.

For granular soils, modulus values (E) have been empirically derived by Meyerhof [3.22], Schmertman [3.27], and others from static cone resistance tests. The relationship between modulus and cone point resistance (q_c in tsf) is presently generally accepted as approximately

$$E = 2.0 q_c \tag{3-16}$$

Vesic [3.21] suggests a higher E/q_c ratio and modification of Eq. (3-16) to include relative density as

$$E = 2(1 + D_r{}^2) q_c \tag{3-17}$$

Evaluation of E for granular soils has also been obtained from the results of laboratory tests, generally on reconstituted samples. Considering that modulus values of granular soils are stress-dependent and the stress-strain relationship is nonlinear, Chang and Duncan [3.28] have shown the tangent modulus (E_t) to be related to the principal stresses (σ_3 and σ_1) and the initial tangent modulus (E_i) by

$$E_t = \left[1 - \frac{R_f(1 - \sin \phi)(\sigma_1 - \sigma_3)}{2\sigma_3 \sin \phi} \right]^2 E_i \tag{3-18}$$

where R_f is the ratio between the compressive strength determined by laboratory tests and the value of the asymptotic stress difference of a hyperbolic stress-strain curve representation of the test data. The E_i

of this relationship after Janbu [3.29] is given as

$$E_i = K_e p_a \left(\frac{\sigma_3}{p_a}\right)^n \tag{3-19}$$

where p_a is the atmospheric pressure in the same units as σ_3. Values of ϕ, R_f, K_e, and n can be determined from the results of conventional triaxial compression tests. Typical values of these parameters, reported by Kulhawy et al. [3.30], are shown in Table 3-3. A similar approach for the laboratory evaluation of the cohesive soil modulus is also demonstrated by Chang and Duncan.

According to Bjerrum [3.31], the undrained E of normally consolidated clay samples is of the order of 250 to 500 times the undrained shear strength (s_u) as determined by uniaxial compression tests. Ladd [3.32] shows the E vs. s_u relationship of overconsolidated clays to be particularly dependent on stress history and the method of testing, as well as being quite sensitive to sample disturbance. Thus, determination of E from routine laboratory shear tests on samples of cohesive soils may prove quite misleading. In the few cases where E has been calculated from field observations, conventional laboratory determinations are usually shown to significantly underestimate the actual values.

(e) Rock Support The ultimate shaft resistance of cohesive soils, such as expressed by Eq. (3-2), has also been applied to piers drilled into rock. However, the shear strength determined by uniaxial or triaxial compression tests on core specimens representative of the unjointed portions of the rock must be reduced to reflect the effect of fractures and other structural and mineralogical discontinuities influencing the shaft resistance of the *in-situ* rock mass much as the undrained shear strength of stiff clay is reduced to estimate shaft adhesion. Shear strength reduction factors (α_r) to be applied to core-sample test values have not as yet been derived for hard rocks, and the shearing resistance appropriate for

TABLE 3-3 Typical Values of Stress-Strain Parameters for Clean Sands and Gravels*

Unified soil classification	ϕ, degrees		K_e	n	R_f
	Low confining pressure	High confining pressure			
GW	47	35	500	0.3	0.7
GP	46	38	1,800	0.3	0.8
SW	50	35	300	0.5	0.7
SP	40	30	1,200	0.5	0.8

* After Kulhawy, Duncan, and Seed (1969).

the design of rock sockets can presently be determined only from load tests conducted so as to measure separately shaft and base resistance.

A sampling of typical design practice indicates that allowable shaft resistance in rock is seldom taken to be more than one-fifth or less than one-tenth the allowable base resistance. The allowable shaft resistance in relatively strong and sound rock is also, in some areas of practice, considered to be controlled by the strength of the shaft concrete. For example, various building codes specify allowable shearing resistances of a socketed shaft to be a function of the compressive strength of concrete, ranging from 0.04 to 0.05 f_c' but not exceeding 200 psi (14 kgf/cm^2).

The base resistance (bearing value) of a pier founded on rock is also dependent on the *in-situ* rock condition, a condition difficult to characterize except by prototype load testing. For indirect evaluation of the compressibility and hence the allowable bearing value of the rock, indices such as rock quality designation (RQD) or fracture frequency (FF) [3.33] and the ratio of the field seismic velocity (V_p) to the laboratory sonic velocity of sound cores (V_L) promise to eventually be as useful as the standard penetration and static cone resistance indices developed for use in soils. For example, using shear (V_s) and compressional (V_p) seismic wave velocities of the *in-situ* rock, together with the longitudinal wave (sonic) velocity of sound core specimens, the unit base resistance vs. settlement relationship can be expressed as

$$\Delta_{bb} = \frac{p_b d_b I_{bb}}{\dfrac{\gamma}{g} V_s^2 \left[\dfrac{3(V_p/V_s)^2 - 4}{(V_p/V_s)^2 - 1} \right] \beta} \tag{3-20}$$

where γ/g is the mass density of the bearing stratum and β is the ratio of the *in-situ* rock modulus (E) to the modulus calculated from seismic velocity (E_{seis}). With additional data, it is probable that the seismic modulus reduction factor (β) can ultimately be developed, as proposed by Deere et al. [3.33], as a function of a rock-quality index. A tentative β vs. rock-quality relationship is shown by Fig. 3-9, which together with the influence factor I_{bb} from Fig. 3-8 enables an estimate of Δ_{bb}. If it could be shown that the ultimate shearing resistance of a given rock mass is directly related to its compressibility, the shear strength reduction factor (α_r) could also be expressed in terms of the aforementioned indices.

(f) Selection of Safety Factors   Safety factors are normally applied to ultimate base and shaft resistance calculated by limit analysis techniques on the basis of strength considerations alone. Usually, F_b is taken to be equal to F_s and is assumed to be at least 2 and more com-

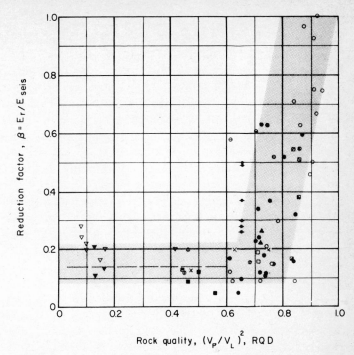

Fig. 3-9 Variation in reduction factor β with rock quality.

monly 2.5. In most instances this approach has resulted in tolerable
settlement behavior. A more rational approach, which considers shaft
and base deformation compatibility, is recommended by Whitaker and
Cooke [3.1]. However, this method depends on an experimentally derived relationship between shaft and base settlement, as shown by
Fig. 3-10.

A simplified approach, developed by Burland et al. [3.19] for London
clays, requires F_b to be 3 and the safety factor for combined shaft
and base load to be at least 2. Also in connection with London clays,
Skempton [3.26] recommends as a first approximation a combined safety

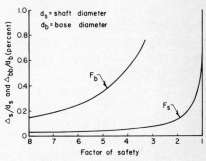

Fig. 3-10 Base and shaft safety
factors vs. respective settle-
ment/diameter ratio. Note that
F_b and F_s are expressed in
terms of the ultimate base
and shaft resistances. [*After
Whitaker and Cooke* (*1966*).]

factor of 2.5 for belled piers where $d_b < 6$ ft (1.8 m) and suggests that the design of larger piers be controlled by settlement considerations.

Unless load-test data from which the pier load distribution can be measured or estimated are available (see "Load Test Interpretation," p. 55), selecting safety factors to quantitatively account for displacement compatibility is not feasible. However, if displacement compatibility is to be weighed in the selection of safety factors for working loads, it is evident that a larger factor should be applied to base resistance than to shaft resistance.

In summary, the obvious limitation of the limit design method results from the inability to rationally consider the shaft and base interaction of a pier without previous knowledge of the load-distribution relationship. It is primarily for this reason that the relatively recent development of displacement compatibility analyses shows the greatest promise as a pier design technique.

3-3.2   Displacement Compatibility Solutions

Provided a realistic pier-to-soil load-transfer relationship can be established, displacement compatibility concepts and elastic, elastoplastic, or finite element analyses can be used to predict—with reasonably good results—the load-settlement behavior of a single pier. If the load transfer from the base and shaft of a pier to the surrounding soil can be expressed in the form of semiempirical (or experimental) relationships between displacement and shaft and base resistance, a load-settlement curve can be readily simulated using a digital computer, as demonstrated by Seed and Reese [3.4], Coyle and Reese [3.37], and others. A method found to yield results approximating observed pier behavior is as follows:

1. Divide the pier into a number of equal segments. The number of segments required for acceptable accuracy can be determined by experiment. From studies using elastic theory, the use of at least ten segments has been suggested by D'Appolonia and Thurman [3.34].

2. If a load-transfer expression for the base has not been developed from field measurements, assume Δ_{bb} and calculate P_b, using an area integration of the Mindlin point-load solution as shown in Fig. 3-8. A nonlinear response can be assumed after the development of some fraction of the predicted ultimate resistance and can be simulated as shown in Fig. 3-11.

3. Assume the midpoint deflection of the lowest segment to be equivalent to the base deflection plus an assumed elastic deflection (Δ_s) of the segment. Find the compatible shearing resistance from the appropriate load-transfer expression (such as shown in Fig. 3-11) and calculate the vertical load (ΔP_s) carried by the segment.

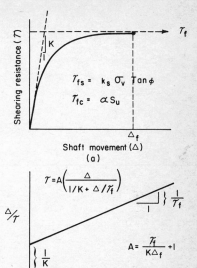

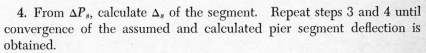

Fig. 3-11 (a) Nonlinear load-transfer approximation by two-constant hyperbola. (b) Transformed hyperbolic function.

4. From ΔP_s, calculate Δ_s of the segment. Repeat steps 3 and 4 until convergence of the assumed and calculated pier segment deflection is obtained.

5. Calculate shaft loads acting on each succeeding segment, progressing to the top of the shaft.

6. Replace the calculated segment loads (ΔP_s) with equivalent point or ring loads acting at the midpoint of the segment and calculate the additional base deflection due to these loads. Influence factors for segment loads derived from the Mindlin point-load solution are given in Fig 3-7.

7. Repeat steps 2 through 5 until compatibility between load and deflection is achieved at the midpoint of each segment.

8. Repeat steps 2 through 7 for successively greater base displacements and construct a load-settlement curve.

The foregoing reiterative operations permit load-transfer vs. movement of a pier segment to be conveniently represented as a nonlinear function, simulating actual field behavior. Closed-form elastic and elastoplastic (elastic-slip) solutions employing displacement compatibility concepts have been developed by several investigators, including D'Appolonia and Thurman [3.34], Salas and Belzunce [3.35], Nair [3.36], and Poulos and Mattes [3.7, 3.8]. These solutions, however, are dependent upon the ususally unrealistic assumption of a uniform and linearly elastic soil mass. An example of a solution for the settlement of the top of a compressible pier embedded in a nonslipping semi-infinite elastic mass is included as Fig. 3-12 as a function of the k_p and L/d parameters of the pier.

(a) Load-transfer Relationships   The suitability of the foregoing type of displacement compatibility solution depends primarily on the suitability of the load-transfer expression. As suggested previously, this relationship can be established with some degree of confidence only by load tests on properly instrumented piers. Even under these conditions, it is often impossible to predict the long-term load-transfer effects of materials with pronounced time-dependent properties.

Load-transfer relationships for pile design, based on analysis of both pile and special laboratory tests, have been proposed by Coyle and Reese [3.37] for clays and by Coyle and Sulaiman [3.38] for sands. Both methods apparently show a reasonable agreement between the predicted and observed behavior of test piles and may possibly be modified for use with drilled piers. Although, to the authors' knowledge, reliable predictions of pier-to-soil load transfer entirely from laboratory test data have not as yet been established, this approach appears promising, particularly for tests (such as simple shear) which can approximate deformation conditions at the pier-soil interface.

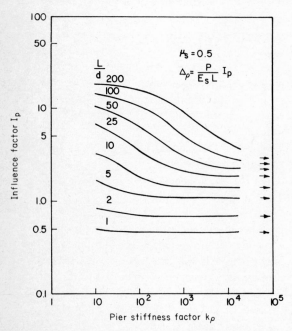

Fig. 3-12 Settlement of top of compressible pier in semi-infinite elastic mass.

(b) Limitations and Application The expression for base resistance vs. base deflection used in the suggested compatibility solution is predicated exclusively on elastic behavior. To better simulate the true load-settlement relationship for a pier at loads beyond the elastic range, an assumption can be made concerning the point at which a nonlinear response is to be initiated, as well as the magnitude of settlement required to develop full base resistance. Moreover, the elastic solutions used for base displacement analysis predict the development of tensile stresses in the material just above the base. Such stresses could not normally be sustained by the bearing materials unless countered by gravity stresses.

With the increasing availability of the digital computer, it has become feasible to use displacement compatibility methods incorporating nonlinear, finite element solutions, which may prove to be far more rigorous than the methods outlined herein. However, it has been shown that, regardless of the rigor of the mathematical model established for computer analysis, all techniques suffer from much the same limitation as limit analysis, that of establishing an appropriate material characterization. Considering the difficulty in establishing the normal variation of material properties as well as the influence of construction operations on these properties, it is unlikely—at least within the near future—that displacement compatibility or other design methods will supplant to a significant degree the widely used limit design technique. In any event, the more rigorous analytic techniques provide the designer with valuable insight into pier load vs. settlement relationships and eventually, upon development of reliable material characterizations, will provide him with a far more powerful pier design method than is currently used in practice.

3-3.3 Load Test Interpretation

Upon completion of a load test, an interpretation must be made to arrive at the design capacity of the test piers. There are many interpretative procedures presently in use, most of which are applied to the load vs. gross or net (nonrecoverable) settlement curve. To predict design capacity, these procedures can usually be generalized into one or a combination of the following methods:

1. A reduction of the measured or projected ultimate (failure) load, usually by a safety factor between 2 and 3

2. The test load corresponding to a limiting gross settlement (usually between $\frac{1}{2}$ and 1 in.) divided by a safety factor (usually 1.5 to 2)

3. The test load corresponding to a limiting rate of net or gross settlement (usually 0.01 or 0.03 in./ton) divided by a suitable safety factor

4. The test load corresponding to a limiting net settlement (usually ¼ in. at the working load (P) and ½ in. at 1.5P)

None of these methods enables a separation of the pier loading into shaft and base resistance, although this interpretation is necessary for an understanding of how the load is transferred to the soil and for the rational selection of safety factors. An approximate division of load between the shaft and base resistance at design load is particularly important where a significant portion of the pier load will be developed by shaft resistance, and is especially needed if development of downdrag (a reversal of shaft resistance) is anticipated. Therefore, the following discussion is confined to consideration of interpretative techniques which enable estimates of load division and determination of rational safety factors.

(a) Instrumented Pier Tests Instrumentation of piers by means of strain rods, strain gauges, and load cells has been used to enable indirect (by strain instruments) and direct (by load cells) measurements of pier load distribution with depth, or simply the amount of base load. Excellent references to the more conventional instrumentation techniques for piers and piles are given by Snow [3.39] and Mansur and Hunter [3.40].

Interpretation of the average load (P) carried by a segment of a pier (ΔL) from measurement of the deflection (R) of two strain rods can be expressed, from Fig. 3-13, as

$$P = a_s E_c \frac{R_1 - R_2}{\Delta L} \qquad [(3\text{-}21)$$

The base load and base deformation can therefore be simply interpreted from the movement of the two strain rods (telltales), one fixed at the tip of the pier and one fixed at a distance above the tip sufficient to yield a measurable differential movement between the freestanding rods. A single strain gauge or load cell can also be used to monitor base load, but it is more susceptible to malfunction and will not record the base deflection. Where complete or even partial load-profile instrumentation is not practical, a single telltale fixed at the base to record both base movement and elastic shortening of the shaft is recommended in lieu of interpretative techniques which involve time-consuming cycling of the test load in an attempt to estimate the elastic or nonrecoverable pier movement. A telltale assembly for this kind of measurement is illustrated in Fig. 3-13.

(b) Interpretation of Load-settlement Curve Van Weele [3.41], Hanna [3.42], and others have proposed load-settlement curve interpretations to indirectly evaluate the shaft (P_s) and base (P_b) load division. The Van Weele method uses frequent cycling of the test load

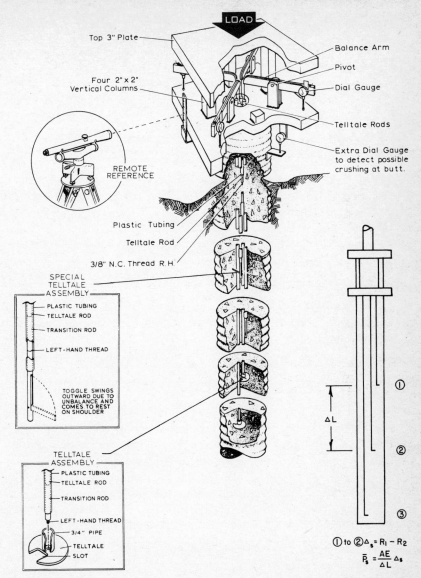

Fig. 3-13 Strain rod (telltale) instrumentation to monitor load distribution in a concrete pile. This instrumentation is readily adaptable to drilled pier testing. (*Raymond International, Inc.*)

to define load vs. rebound of the pier top, from which the shaft load is estimated. Hanna suggests that the point of the load-settlement curve corresponding to the initiation of base loading can be approximated by the point of tangency to the curve of a line drawn at the slope of $a_s E_c / L$, as shown by Fig. 3-14*b*.

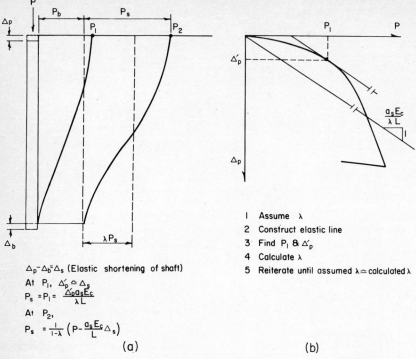

Fig. 3-14 Tangent line method of estimating P_s at inception of base loading.

Provided the elastic deformation of a pier shaft can be measured or estimated, P_s and P_b can be simply derived from consideration of the shaft load distribution. As shown by Fig. 3-14a, the elastic shortening of the shaft can be expressed in terms of the average compressive load in the shaft $\bar{P}_s = \lambda P_s$, where λ is the shaft load influence factor. At the inception of base loading, all the load is carried by the shaft ($P = P_s$), and the settlement of the top of the pier (Δ_p) is primarily the elastic shortening of the shaft (Δ_s); then P_s can be simply expressed as

$$P_s = \frac{a_s E_c \Delta_p}{\lambda L} \tag{3-22}$$

The corresponding point on the P vs. Δ_p curve can be located by a tangent to the curve at a slope of $a_s E_c / \lambda L$. Although λ would be expected to be somewhat time-dependent and to vary for shaft loads less than ultimate, it is believed that for piers in firm-to-hard clays, in soft rocks, and in granular soils, λ will usually range between $\frac{1}{3}$ and $\frac{2}{3}$.

Equation (3-22) does not necessarily yield the ultimate P_s, as additional shaft load may develop concurrent with the mobilization of base load, depending on the L/d_s of pier and the relative stiffness of the mate-

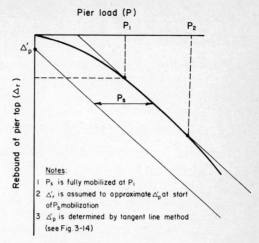

Fig. 3-15 Modified Van Weele method of estimating shaft load.

rial around the shaft and below the base, as well as the relative stiffness of the shaft and the supporting soils. It is also noted that the tangent line method is appropriate only to long, relatively "compressible" piers which develop appreciable shaft resistance prior to initiation of base loading.

Extending the foregoing analysis to within the range of base loading, P_s can be readily derived from Fig. 3-14 as

$$P_s = \frac{1}{1 - \lambda}\left(P - \frac{a_s E_c \Delta_s}{L}\right) \qquad (3\text{-}23)$$

The elastic shortening of the shaft (Δ_s) and base settlement (Δ_b) are directly determined by a single telltale fixed at the level of the pier base. By this means P vs. Δ_b can also be measured, and the point at which base load is initiated can be directly determined and the corresponding λ calculated from Eq. (3-22).

If Δ_s is not directly measured by a base telltale, a combination of the Van Weele and tangent line methods can be used with test-load cycling to separate base and shaft load. This technique, demonstrated by Fig. 3-15, includes the following steps.

1. Assume λ' and draw a tangent to the P vs. Δ_p curve at a slope of $\lambda' L / a_s E_c$.

2. Calculate $\lambda = \Delta'_p a_s E_c / p' L$ from Δ_p and P corresponding to the point of tangency.

3. Reiterate steps 1 and 2 until λ' equals λ.

4. Assume Δ'_p to be the point at which base load just begins to be mobilized and assume Δ'_p = elastic rebound of pier top (Δ'_r). Note that Δ'_p of short "stiff" piers would approach zero.

5. Plot Δ_r vs. P and determine the slope of the linear portion of the curve (the portion of the curve between P_1 and P_2 on Fig. 3-15).

6. Estimate ultimate P_s as the difference between the ordinates of the P vs. Δ_r curve and a line through $P = 0$, $\Delta_r = \Delta'_r$ drawn parallel to the linear portion of this curve, as shown on Fig. 3-15.

It is noted that P_s derived by the modified Van Weele method will tend to be underestimated because of residual stresses in the shaft after unloading.

If Δ_s is directly measured, P_s can also be graphically interpolated from the P vs. Δ_s curve in accordance with Eq. (3-23). The quantity $(P - a_s E_c \Delta_s / L)$ becomes the horizontal dimension (z) between the P vs. Δ_s curve and a straight line through the origin drawn at a slope of $a_s E_c / L$. The shaft load is then simply $z/(1 - \lambda)$, and λ is calculated as $\Delta'_s a_s E_c / PL$, where Δ'_s is the shaft shortening at the observed inception of base loading.

3-3.4 Group Action

Subsurface conditions favoring the use of axially loaded drilled piers in groups would usually be limited in practice to thick deposits of stiff clay (such as London clay) or soft rocks capable of sustaining only moderate base loads. When the center-to-center spacing of such piers is greater than eight shaft diameters (d_s), group piers are normally designed as single units and the consolidation settlement of the group is checked [3.43]. At closer spacings, the load capacity of a group (P_g) in cohesive soils is known to be less than single pier capacity (P) multiplied by the number of piers in the group (n_g); i.e., the efficiency (E_f) of the group is less than 1. Correspondingly, group capacity can be expressed as

$$P_g = E_f n_g P \tag{3-24}$$

To estimate E_f for relatively uniform deposits, Tomlinson [3.44] recommends a linear interpolation from 0.7 at $3d_s$ to 1.0 at $8d_s$. In cohesionless soils, E_f is generally assumed to be 1.0, although higher efficiencies have been noted, depending on the pier spacing.

Because the foregoing simplified evaluation of E_f is based primarily on model tests in carefully prepared uniform clays, and because it has been found that E_f is somewhat dependent on group size, caution is required in application. Further, the group efficiency analyses consider only shear capacity, whereas settlement may control design and must be considered in all instances of piers drilled into cohesive materials.

At a spacing less than $3d_s$, a pier group of width W_g and breadth B_g may fail in shear as a single block. Under these circumstances, the ultimate capacity of a pier group of length L_g can be expressed for cohesive soils as:

$$P_g = 2L_g(W_g + B_g)s_u + 1.3N_cW_gB_g \qquad (3\text{-}25)$$

The width and breadth of the group are measured as the average dimensions of a rectangle circumscribing the peripheral piers, and the *in-situ* shear strength around the group periphery may be assumed to be reduced by construction influence.

A second approach to group evaluation uses the ratio of group settlement to the settlement of a single pier as a measure of group efficiency. This approach has been developed by Skempton et al. [3.45] for piles in sands by interpreting field measurements. The settlement efficiency of a pile or pier group has also been investigated using theoretical considerations, usually based on elastic theory. The comprehensive work of Poulos [3.46] is particularly noteworthy and permits an analysis which considers several of the parameters known to influence group action.

The use of battered piers in groups, although an infrequent occurrence, may be applicable to the resistance of large horizontal loads which cannot be sustained by the lateral reaction of vertical piers. Under these circumstances, it is appropriate to consider the combined resistance of vertical and battered piers, as proposed by Hrennikoff [3.47]. An excellent synopsis of this method of analysis, best conducted with the aid of a digital computer, is presented by Bowles [3.48]. Bowles also describes a simple, approximate method of batter pier analysis which assumes group piers to carry only axial load.

3-4 BEHAVIOR OF LATERALLY LOADED PIERS

The behavior of a laterally loaded pier is dependent not only on the many parameters listed in Art. 3-2, which control the load response of single, axially loaded piers, but also to a large extent on the degree of fixity imposed at the top of the pier by the supported structure. Also, in comparison to axial load response, behavior under lateral load is to a much larger degree controlled by the flexural stiffness of the pier relative to the stiffness of the materials surrounding the upper portion of the shaft. The load vs. deflection characteristics of a "rigid" pier are therefore quite different from those of an "elastic" pier. A theoretical demarcation between "rigid" and "elastic" pier behavior has been developed from the analysis of a beam bearing on an elastic subgrade, as will be subsequently considered.

Most analyses of piers subject to lateral loads and/or overturning moment are concerned with behavior at loads of one-half to one-third of the ultimate soil resistance or of the ultimate structural resistance of the pier and involve predictions of deflection, bending moment, shear, and soil reaction.   Such predictions cannot be obtained from limit (ultimate load) analysis techniques.   Consequently, theoretical analyses over the past 25 years have concentrated on the development of concepts based on displacement compatibility considerations.   The resultant design techniques have become well developed, and constitute the present state of the art for analysis of laterally loaded piles and, with some limitations, of laterally loaded piers.

3-4.1   Elastic Pier Analysis

The behavior of an elastic pier or pile is considered to be closely related to the behavior of a long beam on an elastic subgrade.   Correspondingly, elastic pier behavior has been expressed as a function of depth (x), lateral deflection (y), axial load (Q), flexural stiffness (EI), and soil modulus (K) in accordance with Eq. (3-26).

$$EI\frac{d^4y}{dx^4} + Q\frac{d^2y}{dx^2} + Ky = 0 \qquad (3\text{-}26)$$

As the buckling load has been shown [3.49] to be well in excess of the axial load likely to be applied, the second term of Eq. (3-26) is usually ignored, and a simplified expression for a beam on an elastic subgrade or for an elastic pile, differing only with respect to boundary conditions, is given as

$$EI\frac{d^4y}{dx^4} + Ky = 0 \qquad (3\text{-}27)$$

Solutions of Eq. (3-25), facilitated by finite difference methods [3.50], have been developed for various modulus functions ranging from a linear elastic assumption (constant k) to a variation of modulus in both the x and y directions.   A typical solution of Eq. (3-27) for soil reaction, moment, and deflection is shown by Fig. 3-16.   It is noted that all the solutions required for structural analysis can be obtained by successive integration and/or differentiation from any one solution.   References to published elastic pier/pile solutions applicable to various soil modulus variations are listed in Table 3-4.   Poulos [3.51] has also derived a solution for laterally loaded piers and piles using elastic theory, independent of the soil modulus concept.

Reese and Matlock [3.53] have derived elastic pier solutions that account for nonlinear soil behavior by repeated application of elastic theory, using a series of iterative approximations to obtain a variation of

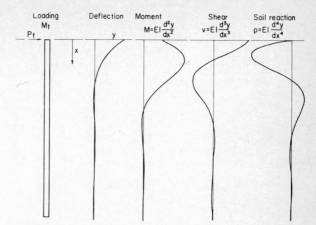

Fig. 3-16 Elastic pier behavior. [*After Matlock and Reese* (*1962*).]

TABLE 3-4 Elastic Pier Solutions

Modulus variation with depth	Solution reference
Constant...............................	Grandholm, Hetenyi [3.49, 3.52]
Linear increase.........................	Reese and Matlock, Reese and Ginzbarg* [3.53, 3.54]
Exponential increase....................	Palmer and Thomson, Matlock and Reese [3.55, 3.56]
Polynomial form of increase.............	Matlock and Reese [3.56]
Constant, two layer.....................	Davisson and Gill [3.57]

* For "step-tapered" cross section.

soil modulus with depth compatible with the assumed modulus function. This method, using actual stress vs. deflection relationships (called *p-y* curves) established from field or laboratory tests, involves nondimensional coefficients and is believed to presently represent the most plausible and flexible design approach.

3-4.2 Rigid Pier Analysis

If a pier is sufficiently stiff to displace (rotate) under lateral load without appreciable distortion from its axis, the lateral displacement at any depth may be described with reference to a center of rotation as shown by Fig. 3-17. This condition simplifies the displacement compatibility analysis and has generated a proliferation of "rigid pole" solutions. An excellent review of rigid pole literature has been published by Davisson and Prakash [3.58].

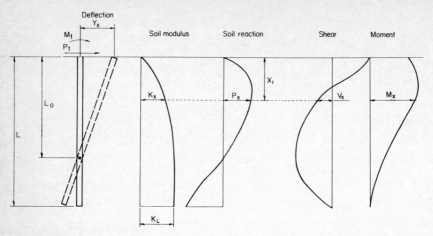

Fig. 3-17 Rigid pier behavior.

The lateral load-deflection response of a rigid pier may be significantly influenced by the moment resistance of the base and by side-shear effects on the shaft not considered in most of the published rigid pole solutions. The effect of base conditions has been demonstrated by load tests on a short underreamed pier and on a short socketed pier, as reported by Davisson and Salley [3.59]; both types of pier yielded significantly higher resistances to lateral load than were predicted by pole (rigid pier) theory. When a pier can be considered to be essentially rigid in comparison with the surrounding subsoils and is not socketed or underreamed, it is probable that pole theory is reasonably appropriate, provided the embedded length of the pier is greater than about four to five diameters, as is the usual case for drilled piers. A solution for a short rectangular pier which includes base and side-shear effects has been derived by Roscoe and Schofield [3.60] for sands. An empirical solution based on the results of model tests in various soil types has been developed by Ivey et al. [3.61].

3-4.3   Elastic vs. Rigid Pier Behavior

Solutions of Eq. (3-27) indicate that the demarcation between elastic and rigid pier behavior can be determined in terms of a relative stiffness factor which expresses a relation between soil stiffness and pier flexural stiffness and is dependent upon the soil modulus function assumed. For a constant soil modulus assumption, the stiffness factor (R) can be described in accordance with Eq. (3-28), where the dimensions of K and R are in units of force/length² and length, respectively.

$$R = \sqrt[4]{\frac{EI}{K}} \qquad (3\text{-}28)$$

For a soil modulus increasing linearly with depth, the stiffness factor (T) (dimension: length) is expressed in accordance with Eq. (3-29)

$$T = \sqrt[5]{\frac{EI}{n_h}} \qquad n_h \text{ see page 68} \qquad (3\text{-}29)$$

where n_h (dimension: force/length^3) is the coefficient of horizontal subgrade reaction.

For the constant-modulus, free-head solution, it is generally accepted that rigid pier behavior can be assumed when the pier length L is less than $2R$, and that a pier behaves as an infinitely long member when $L \geq 3.5R$. For the linearly-increasing–modulus, free-head solution, rigid behavior is assumed when $L \leq 2R$ and the infinite-length solution is approximated when $L \geq 4T$. The theoretical influence of pierhead fixity on the demarcation between rigid and infinite-length behavior and on the amount of ground line deflection is summarized in Table 3-5.

TABLE 3-5　Head Fixity vs. Deflection (y_0)

Pier	Soil modulus			
	Linearly increasing		Constant	
	Free pier head	Fixed pier head	Free pier head	Fixed pier head
Elastic*....	$L \geq 4T$ $$y_0 = \frac{2.44P_tT^3 + 1.62M_tT^2}{EI}$$	$L \geq 4T$ $$y_0 = \frac{0.93P_tT^3}{EI}$$	$L \geq 3.5R$ $$y_0 = \frac{1.41P_tR^3 + M_tR^2}{EI}$$	$L \geq 2R$ $$y_0 = \frac{0.71P_tR^3}{EI}$$
Rigid.......	$L \leq 2T$ $$y_0 = \frac{18P_t + 24M_t/L}{L^2n_h}$$	$L \leq 2T$ $$y_0 = \frac{2P_t}{L^2n_h}$$	$L \leq 2R$ $$y_0 = \frac{4P_t + 6M_t/L}{KL}$$	$L \leq 0.7R$ $$y_0 = \frac{P_t}{KL}$$

* "Infinite length" behavior.

3-4.4　Soil Modulus

As indicated previously, the soil modulus variation can be assumed to be constant with depth, to linearly or exponentially increase with depth, or to vary with depth in some other continuous fashion. Further, the soil at a given depth can be assumed to be linearly elastic or, more realistically, stress-dependent. However, it has been shown that calculated bending moments and deflections are not very sensitive to changes in K, and that it is usually sufficient to assume simple forms of variation with depth and to introduce refinements by adjustments of soil parameters. The following discussion of soil modulus is therefore primarily directed toward application to the more simplified modulus variations.

(a) Modulus Constant with Depth   For the constant-modulus variation, usually taken as applicable only to preconsolidated clays, the ratio of deflection (y_x) to unit soil reaction (p_x), i.e., $k = p_x/y_x$, is commonly defined as the coefficient of subgrade reaction (k) in units of force/length^3 and is used interchangeably with either beam or pier theory.   It is also appropriate to express soil reaction (p_x) in units of force/length (the reactive force per unit length of pier), and the corresponding soil modulus in units of force/length^2, as $K = kd_s$ where d_s is the pier diameter.   Terzaghi [3.62] has demonstrated that it is convenient to relate k to field load tests on a small (usually 1-ft-square) plate $(k_1,)$ and recommends

$$k = \frac{k_1}{1.5d_s} \tag{3-30}$$

where d_s is the width of the embedded portion of a pier.   Since $K = d_s k$, soil modulus (K) expressed in terms of the coefficient of subgrade reaction (k_1) is numerically independent of the width of the pier.   For short (rigid) piers K is dependent only on the soil properties and can be expressed by Eq. (3-30), or more rigorously, as proposed by Broms [3.63], by assuming the deflection to be a superposition of the rotation and translation modes of the pier.

The experimental findings of Vesic [3.64] indicate that for piers with $L \geq 3R$, K is significantly influenced by the flexural stiffness (EI) and the diameter of the member, and can be expressed by Eqs. (3-31) and (3-32) in terms of either K_1 or the modulus of soil deformation (E_s).

$$K = 0.52K_1 \sqrt[12]{\frac{K_1 d_s^4}{EI}} \tag{3-31}$$

$$K = 0.65 \sqrt[12]{\frac{E_s d_s^4}{EI}} \frac{E_s}{1 - \mu^2} \tag{3-32}$$

Note that E_s can be evaluated from the results of laboratory tests as suggested by Francis [3.65] and by Davis and Poulos [3.66].   Francis also recommends that K values determined from Eqs. (3-31) and (3-32) be doubled to account for complete embedment, not considered in the "beam on elastic subgrade" derivation, and observed that the term $\sqrt[12]{d_s^4/I}$ is 1.28 for a solid circular section; thus,

$$K = \frac{1.60 E_s^{13/12}}{E^{1/12}(1 - \mu^2)} \tag{3-33}$$

Values of k_1 from tests using 1-ft-square bearing plates on a preconsolidated clay subgrade are summarized after Terzaghi [3.62] in Table 3-6.   Based on reported correlations between undrained shear strength (s_u) and deformation modulus (E_s) [3.63], these values, which are about 50 times the undrained shear strength, appear quite conservative and undoubtedly allow for long-term reductions due to consolidation and creep. Allowing for a two-thirds reduction, more recent data suggest $k_1 \simeq 67\ s_u$ as a conservative approximation.

TABLE 3-6 k_1 **Values for Precompressed Clay***
(Tons per cu ft or 0.032 kg/cm^3)

Consistency:	Stiff	Very stiff	Hard
s_u in tsf (kgf/cm^2)............	1–2	2–4	Greater than 4
k_1 range.....................	50–100	100–200	Greater than 200
k_1 recommended..............	75	150	300

* After Terzaghi.

Both load-test results and theoretical analysis indicate that the soil properties from the surface down to a depth equivalent to about 2.5R (free-head) or 1.5R (fixed-head) essentially control pier behavior. Therefore, a more detailed consideration must be given to the selection of K in this interval and to the influence of seasonal changes on soil properties.

To account for the influence of long-term consolidation of laterally loaded stiff-to-very-stiff clays, Broms [3.63] suggests that K should be taken as one-half to one-fourth of the value applicable to undrained conditions. For normally consolidated clays, the suggested reduced value is $\frac{1}{3}$ to $\frac{1}{6}$ K. The influence of repetitive loading is also to reduce K and to increase free-head deflections as much as 100 percent, depending on the magnitude of lateral load and the number of load cycles. For example, Alizadeh [3.67], reporting on the results of tests on timber piles in firm-to-soft clays, shows a significant increase in deflection after about 20 cycles of a 5-ton (4,550-kg) lateral load and notes a 57 percent increase in deflection after 20 cycles of a 10-ton (9,100-kg) lateral load.

(b) Modulus Increasing Linearly with Depth The measured moment variation of test piles in several instances has shown best agreement with an analysis that uses an assumption of modulus increasing linearly with depth. Presently this assumption is usually accepted as the best approximation for granular soils, normally consolidated clays, and other materials whose modulus can be assumed to approximately increase with depth. The linearly increasing soil modulus (k) can be expressed in terms of a constant of subgrade reaction (n_h) and depth (x) as

$$K = n_h x \tag{3-34}$$

where n_h, in units of force/length^3, can be determined from plate load tests using a small, usually 1-ft-square, rigid bearing plate. In the absence of specific test information, Terzaghi [3.62] recommends that n_h be deter-

TABLE 3-7 n_h **for Moist and Submerged Sands** *
(Tons/ft^3 or 0.032 kg/cm^3)

Sand	Relative density		
	Loose	Medium	Dense
Dry or moist............	$n_h = 7$	21	56
Submerged..............	$n_h = 4$	14	34

* After Terzaghi.

mined for sands as a function of relative density in accordance with Table 3-7.

In soft normally consolidated clays, n_h values of 1 to 2 tons/ft^3 (0.03 to 0.06 kg/cm^3) have been observed. Values as low as 0.5 tons/ft^3 have been reported for soft organic silts.

Loose and medium-dense sands, being particularly susceptible to densification under load repetition, have been found to undergo a significant reduction in n_h under cyclic loading. Broms [3.68] suggests that a reduced n_h be selected for cases of cyclic loading on the basis of initial relative density, recommending $\frac{1}{4}n_h$ for "low" relative density and $\frac{1}{2}n_h$ for "high" relative density sands.

(c) Nonlinear-modulus Function In addition to varying with depth, soil modulus would be expected to vary with the level of stress (and deflection) imposed at a given depth. This variation in terms of n_h is demonstrated by Fig. 3-18, which typifies the results of an analysis [3.59] of lateral load tests on long (elastic) piers.

As previously indicated, if the p-y relationship can be described at successive depths along a pier, a solution can be obtained for any given continuous modulus variation with depth by using a reiterative procedure involving interpolation of successive secant moduli directly from the curves. Although derivation of reliable p-y curves for most situations is dependent on analysis of field load tests, McClelland and Focht [3.69] have suggested, on the basis of a similarity between triaxial compression test results and p-y curves derived from field tests, that laboratory tests can be used to simulate p-y curves for soft-to-firm clays. Figure 3-19 shows the p-y configuration for typical lateral load test data. The curve shape also suggests the applicability of mathematical simulation by a two-constant hyperbolic transformation, similar to that illustrated by Fig. 3-11.

3-4.5 Ultimate Lateral Resistance

Failure of a laterally loaded pier will occur when either the available soil resistance or the flexural resistance of the pier is exceeded, the

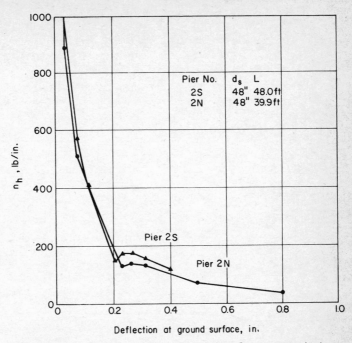

Fig. 3-18 Coefficient of horizontal subgrade reaction (n_h) vs. deflection at ground surface. [*After Davisson and Salley (1968).*]

failure of elastic piers occurring as a flexural overstress of the shaft. The various idealized failure mechanisms, as influenced by head fixity conditions, are shown by Fig. 3-20. In the following discussion, the flexural failure mechanism is assumed, after Broms [3.70], to be analogous to a "plastic hinge" in the pier shaft which is capable of redistributing bending moment and of sustaining imposed shear.

(a) Ultimate Soil Resistance A rigorous analysis of the lateral resistance mobilized upon development of the ultimate soil resistance of a laterally loaded pier is a complex problem in displacement compatibil-

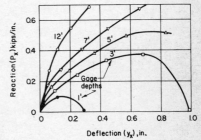

Fig. 3-19 Soil reaction vs. deflection of laterally loaded pier (pile). [*After McClelland and Focht (1956).*]

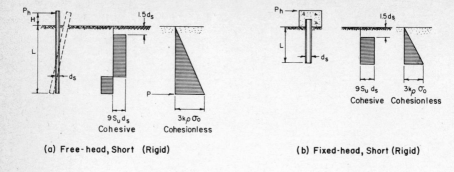

(a) Free-head, Short (Rigid)   (b) Fixed-head, Short (Rigid)

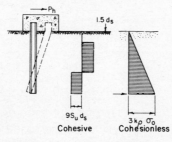

(c) Fixed-head, Intermediate

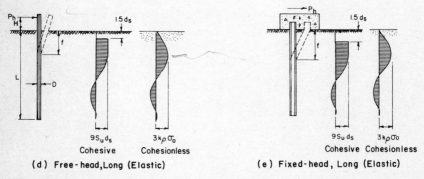

(d) Free-head, Long (Elastic)   (e) Fixed-head, Long (Elastic)

Fig. 3-20   Failure mechanism of free- and fixed-head piers.   [*After Broms (1964)*.]

ity, involving a progressive failure mechanism and soil reaction intensities well beyond the range where elastic soil behavior can be assumed. Considering the analytic difficulties and the inexactness involved in expressing soil behavior, it is believed that the following approximate methods of analysis are appropriate for both cohesive and cohesionless soils.

The lateral bearing capacity (q_h) at a given depth along a pier or pile in cohesionless soil can be expressed by Eq. (3-35) as a function of overburden pressure (σ_0) and a lateral bearing capacity factor (N_{qh}).

$$q_h = N_{qh}\sigma_0 \qquad (3\text{-}35)$$

Hansen [3.71] has derived, from theoretical considerations, values of N_{qh} which are dependent only on depth ratio (x/d_s), pier diameter (d_s), and friction angle (ϕ). From Fig. 3-21, N_{qh}, within the depth effectively controlling pier behavior (typically three to five diameters), is seen to vary between 8 and 10 for loose sand $(\phi = 30°)$ and between 12 and 15 for medium-dense sand $(\phi = 35°)$. Various interpretations of full-scale and model tests [3.58] have led to the deduction that q_h/σ_0 may range from two to four times the coefficient of passive earth pressure (k_p). Accordingly, Broms has proposed

$$q_h = 3k_p\sigma_0 \qquad (3\text{-}36)$$

in which $3k_p$ is in reasonably good agreement with the average N_{qh} value

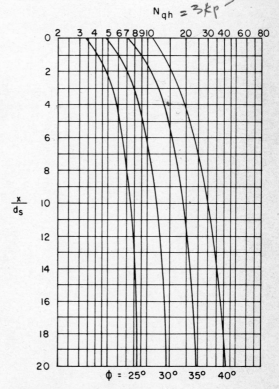

Fig. 3-21 Lateral bearing capacity factor for granular (cohesionless) soil. [After Brinch Hansen (1961).]

for medium-dense sands if taken within a depth equivalent to three to four pier diameters.

The ultimate resistance of cohesive soils in terms of undrained shear strength (s_u) is generally accepted to increase from $2s_u$ at the surface to a maximum of 8 to $11s_u$ at a depth equivalent to about three pier diameters, and not to be affected by the roughness or shape of the pier shaft. Following Broms' recommendations [3.70], the ultimate unit resistance of cohesive soil below a depth of $1.5d_s$ can be expressed as

$$q_h = 9s_u \tag{3-37}$$

It has been noted that repeated loading may ultimately effect a reduction in s_u and significantly reduce the ultimate lateral resistance of a pier in cohesive soils.

(b) Rigid Piers The ultimate lateral resistance (P_u) of both fixed- and free-head piers can be approximated by assuming that the ultimate soil resistance can be calculated from Eqs. (3-36) and (3-37) and is distributed along the pier as shown by Fig. 3-20. From static equilibrium equations, the ultimate lateral resistance of rigid piers in cohesionless soil can be written for a free-head condition as

$$P_u = \frac{0.5d_s L^3 k_p \gamma}{H + L} \tag{3-38}$$

where L is the embedded length of the pier and H is the distance of the resultant lateral load above the ground surface. For fixed-head conditions, the ultimate lateral resistance of rigid piers in cohesionless soils is also dependent on embedment depth and can be written as

$$P_u = 1.5\gamma d_s L^2 k_p \tag{3-39}$$

The ultimate lateral resistance of rigid, fixed-head piers embedded in cohesive soils can be written, from consideration of static equilibrium, as

$$P_u = 9s_u d_s (L - 1.5d_s) \tag{3-40}$$

For rigid, free-head piers in cohesive soils, the depth to the center of rotation can be expressed as

$$L_0 = \frac{H + \frac{2}{3}L}{2H + L} \tag{3-41}$$

If L' is the embedded pier length and L'_0 is the depth to the center of rotation, both measured from a point $1.5d_s$ below the ground surface, P_u can be derived in terms of L'_0 and L' as

$$P_u = \frac{L_0'^2 - 2L'L_0' + 0.5L'^2}{L' + H + 1.5d_s} 9s_u d_s \tag{3-42}$$

(c) Elastic Piers Because the ultimate lateral resistance of an elastic pier is controlled by the flexural resistance of the shaft, it can be assumed that failure will occur when maximum resisting moment (M_p) is exceeded by the maximum imposed moment and a "plastic hinge" develops at the location of maximum moment. As shown by Fig. 3-20d, the plastic hinge of a free-headed pier develops at the point of maximum positive moment, and the ultimate lateral resistance can be computed by assuming that the ultimate soil resistance will be developed on the portion of the pier above the hinge.

If it is assumed that the behavior of a fixed-head (long) elastic pier is similar to that of a pier of infinite length, the ultimate lateral resistance is materially increased and is developed only after the formation of two plastic hinges—one at the location of maximum negative moment (at the pier cap) and one at the location of maximum positive moment, as shown by Fig. 3-20e.

For piers in cohesionless soils, it can be shown that the depth of the plastic hinge is approximately $0.82(P/\gamma d_s k_p)^{1/2}$ and that, if the maximum positive resisting moment is taken to be $M_p{}^+$, the ultimate lateral resistance (P_u) of a free-head and a fixed-head elastic pier can be written as

Free-head:
$$P_u = \frac{M_p{}^+}{H + 0.54\sqrt{P_u/\gamma d_s k_p}} \qquad (3\text{-}43)$$

Fixed-head:
$$P_u = \frac{2M_p{}^+}{H + 0.54\sqrt{P_u/\gamma d_s k_p}} \qquad (3\text{-}44)$$

Equation (3-44) assumes that the yielding negative and positive moments of fixed-head piers are equivalent (i.e., the pier has a uniform cross section).

For fixed-head piers of uniform section embedded in cohesive soils, the depth below the ground surface (f) to the location of the maximum positive moment is $P_u/9s_u d_s + 1.5d_s$, and the ultimate lateral resistance

$$P_u = \frac{2M_p}{1.5d_s + 0.5f} \qquad (3\text{-}45)$$

For the free-head case, a plastic hinge develops in an elastic pier at a depth of $(f + 1.5d_s)$ below ground surface, and P_u can be expressed as

$$P_u = \frac{M_p{}^+}{H + 1.5d_s + f} \qquad (3\text{-}46)$$

In the case of an elastic pier of intermediate length, the idealized failure mechanism shown by Fig. 3-20c is assumed. The ultimate lateral resistance can therefore be expressed for cohesive and cohesionless soils by substituting M_p^- for M_p^+ in Eqs. (3-43) and (3-46).

Limit design techniques using the foregoing expressions to calculate ultimate lateral resistance must consider the effects of varying environmental conditions, of repetitive loading, and of other mechanisms which would tend to influence the soil parameters k_p and s_u. Further, the influence of group action and the possibility of imposed loads exceeding design values must be considered. The safety factor must also be sufficient to prevent failure by excessive deformation, and for this reason a strict limit design technique is not recommended for structures with a low deflection tolerance.

3-5   DESIGN FOR LATERAL LOAD

From experience primarily gained from lateral load tests on instrumented piles, it has been generally found that reasonably good agreement between predicted and observed behavior is obtained from displacement compatibility solutions involving a soil modulus variation in the form of $K_x = Kx^n$, usually with $n = 1$ or $n = 0$. Within the limitations cited in Art. 3.4, these findings are believed to be conservatively applicable to the analysis of laterally loaded drilled piers, and to be particularly suited to the analysis of elastic piers where $L \geq 2T$ or $\geq 2R$.

A common purpose of each of the solutions discussed herein is the determination of the length and flexural rigidity of a laterally loaded pier which will limit pierhead deflections to tolerable limits and will avoid overstressing the pier or the supporting soil. Methods for calculation of pier deflection, soil reaction, and the other parameters necessary for structural analysis are summarized in the first three sections of this article, pp. 74–78, and analysis of ultimate lateral resistance is considered in the section "Ultimate Load Analysis," p. 78. Some empirical design methods, primarily as reflected by building codes, and comments on the design of pier groups are included in the last two sections.

3-5.1   Elastic Piers—K Increasing
Linearly with Depth

Although solutions for elastic (and rigid) piers can be derived to fit almost any soil modulus variation with depth [3.56], a simple, linearly increasing variation is presently accepted as the most useful and as having the widest application. The complete solution of this case [3.53] for elastic piers and piles is greatly facilitated by the use of nondimensional coefficients for both the free-head and fixed-head cases. Equations for the calculation of shear (v), bending moment (m), slope (s), deflection (y), and soil reaction (p) along a pier are expressed in terms of applied force (P_t), applied moment (M_t), stiffness factor (T), and flexural rigidity (EI). Expressions for m, s, and y are given as

$$m = A_m P_t T + B_m M_t \tag{3-47}$$

$$s = A_s \frac{P_t T^2}{EI} + B_s \frac{M_t T}{EI} \tag{3-48}$$

$$y = A_y \frac{P_t T^3}{EI} + B_y \frac{M_t T^2}{EI} \tag{3-49}$$

where A_y, A_m, B_y, and B_m are coefficients, shown on Fig. 3-22, which are dependent on the depth coefficients $Z = x/T$ and $Z_{max} = L/T$. Table 3-8 contains a tabulation of coefficients applicable to a long pier $(L > 4T)$.

The influence of fixity of the pierhead can be accounted for by equating an expression containing M_t for the slope of the superstructure member to the equation for pier slope at the pierhead (Eq. 3-48) and solving for M_t. For long piers $(L \geq 4T)$, the solution has been greatly facili-

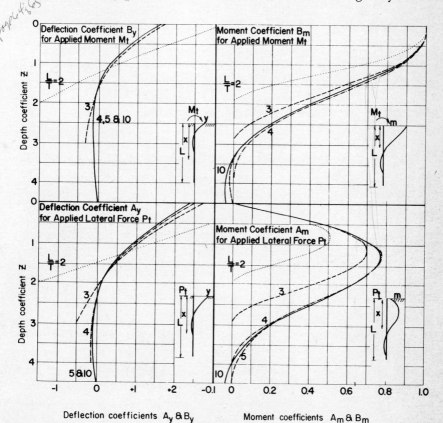

Fig. 3-22 Deflection and moment coefficients for modulus increasing linearly with depth. [*After Matlock and Reese (1956).*]

TABLE 3-8   Coefficients for Long Piles, $K = n_h X$

Z	A_y	A_s	A_m	A_v	A_p	B_y	B_s	B_m	B_v	B_p
0.0	2.435	-1.623	0.000	1.000	0.000	1.623	-1.750	1.000	0.000	0.000
0.1	2.273	-1.618	0.100	0.989	-0.227	1.453	-1.650	1.000	-0.007	-0.145
0.2	2.112	-1.603	0.198	0.956	-0.422	1.293	-1.550	0.999	-0.028	-0.259
0.3	1.952	-1.578	0.291	0.906	-0.586	1.143	-1.450	0.994	-0.058	-0.343
0.4	1.796	-1.545	0.379	0.840	-0.718	1.003	-1.351	0.987	-0.095	-0.401
0.5	1.644	-1.503	0.459	0.764	-0.822	0.873	-1.253	0.976	-0.137	-0.436
0.6	1.496	-1.454	0.532	0.677	-0.897	0.752	-1.156	0.960	-0.181	-0.451
0.7	1.353	-1.397	0.595	0.585	-0.947	0.642	-1.061	0.939	-0.226	-0.449
0.8	1.216	-1.335	0.649	0.489	-0.973	0.540	-0.968	0.914	-0.270	-0.432
0.9	1.086	-1.268	0.693	0.392	-0.977	0.448	-0.878	0.885	-0.312	-0.403
1.0	0.962	-1.197	0.727	0.295	-0.962	0.364	-0.792	0.852	-0.350	-0.364
1.2	0.738	-1.047	0.767	0.109	-0.885	0.223	-0.629	0.775	-0.414	-0.268
1.4	0.544	-0.893	0.772	-0.056	-0.761	0.112	-0.482	0.688	-0.456	-0.157
1.6	0.381	-0.741	0.746	-0.193	-0.609	0.029	-0.354	0.594	-0.477	-0.047
1.8	0.247	-0.596	0.696	-0.298	-0.445	-0.030	-0.245	0.498	-0.476	0.054
2.0	0.142	-0.464	0.628	-0.371	-0.283	-0.070	-0.155	0.404	-0.456	0.140
3.0	-0.075	-0.040	0.225	-0.349	0.226	-0.089	0.057	0.059	-0.213	0.268
4.0	-0.050	0.052	0.000	-0.106	0.201	-0.028	0.049	-0.042	0.017	0.112
5.0	-0.009	0.025	-0.033	0.015	0.046	0.000	-0.011	-0.026	0.029	-0.002

tated for various degrees of head fixity by the use of a modified non-dimensional deflection coefficient (C_y), which has been expressed by Matlock and Reese [3.72] as a function of M_t/P_tT and x/T.

Assuming that p-y curves are developed which express nonlinear soil moduli at selected depths along the pier, the following method of analysis is believed to be the most appropriate for the design of elastic piers subject to horizontal loading:

1. Assume an initial modulus variation such as $K = n_hx$ and compute deflection (y).

2. From p-y curves, determine a soil reaction (p) which is compatible with y and calculate $K = p/y$.

3. Plot K vs. depth (x), fit a straight line (favoring points down to $x = 0.5T$), and determine new n_h values.

4. Recalculate y and n_h, using repeated trials, until the calculated n_h (or T) values of two successive trials are in reasonable agreement.

Until there is more evidence that laboratory test data can be used to simulate actual p-y curves, the nonlinear modulus approach will be dependent on the results of load tests. Reese and Cox [3.73] describe a relatively inexpensive method of establishing p-y curves by recording only the pierhead slope and deflection of uninstrumented piers, and report good agreement between observed and predicted behavior.

3-5.2   Elastic Piers—Constant K

Complete solutions for both applied moment and load have been derived for elastic piles ($L \geq 2R$) with a constant soil modulus by Hetenyi [3.52] and Davisson and Gill [3.57], using nondimensional coefficients. The coefficients applicable to the calculation of deflection (A'_y, B'_y) and bending moment (A'_m, B'_m) by Eqs. (3-51) and (3-52) are included in Fig. 3-23 as a function of dimensionless depth coefficients (Z' and Z'_{max}), where

$$Z' = \frac{x}{R} \quad \text{and} \quad Z'_{max} = \frac{L}{R} \tag{3-50}$$

$$y_x = P_t \frac{R^3}{EI} A'_y + M_t \frac{R^2}{EI} B'_y \tag{3-51}$$

and

$$m_x = P_t R A'_m + M_t B'_m \tag{3-52}$$

Solutions for pier slope (s), shear (v), and soil reaction (P_x) can be obtained from reference 3.52 or by integration or differentiation of the above expressions. The influence of any degree of pierhead fixity can be accounted for, in the same manner as described in the previous section, as a function of the angular restraint provided by the superstructure member restraining rotation of the pierhead.

As described in Art. 3-4, the applicability of the constant K solution is quite limited and is suitable for use only with overconsolidated clays and those rare deposits whose shear strength can be assumed to be relatively constant with depth. Recommendations for selection of soil

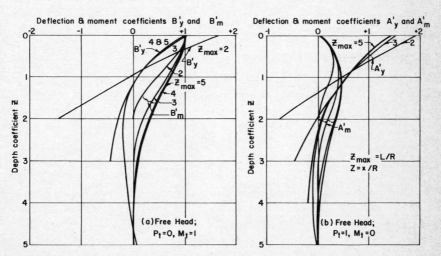

Fig. 3-23 Deflection and moment coefficients for constant modulus. [*After Davisson and Gill* (*1960*).]

modulus as a function of the pier stiffness factor (R) are summarized in Art. 3-4.

3-5.3   Rigid Piers

Since both constant and linearly increasing soil moduli are special cases of the soil modulus variation $K_x = K_L(x/L)^n$, it is convenient to use this expression in a general solution for a free-head rigid pier. Thus for rigid piers ($L \leq 2 \sqrt[n+4]{EI/K}$) equations for pierhead deflection (y_0) and soil reaction (P_x), as adapted from Prakash [3.74], are

$$y_0 = \frac{(n+2)P_t L/L_0}{[(L_0/L)(n+2)/(n+1) - 1]KL} \tag{3-53}$$

$$P_x = K\left(\frac{x}{L}\right)^n y_0 \frac{L_0 - x}{L_0} \tag{3-54}$$

In these equations L_0 is the depth from the surface to the center of rotation, as shown by Fig. 3-17, and

$$\frac{L_0}{L} = \frac{M_t/P_t L + [(n+2)/(n+3)]}{(M_t/P_t L)[(n+2)/(n+1)] + 1} \tag{3-55}$$

The slope of the pier is simply expressed as $s = y_0/L_0$, and the maximum soil reactions occur at $x = [n/(n+1)]L_0$ and at $x = L$. Prakash also presents expressions for y_x and P_x which consider an initial pier rotation and the application of a vertical load.

As K_L is the soil modulus at the base of the pier, it is noted that for granular soils, assuming $n = 1$, $K_L(x/L)^n = n_h x$. Davisson and Prakash [3.58] also report that $n = 0.15$ shows the best agreement with bending moment measurements obtained from a rigid pier embedded in an over-consolidated cohesive soil, recognizing that $n = 0$ is the usual assumption.

3-5.4   Ultimate Load Analysis

For problems in which the design for a pier or group of piers is not controlled by lateral deflection, the limit (ultimate load) analysis approach discussed in Art. 3-4 is applicable. However, where it is necessary to maintain lateral deflection within a specified tolerance, limit design analyses in conjunction with deflection design analyses are appropriate to determine whether or not computed soil reactions are within a range that validates the use of elastic approximations—i.e., whether or not there will be a suitable factor of safety against failure.

(a) Limit Design—Cohesive Soils   To determine the ultimate lateral resistance of both fixed- and free-head piers embedded in cohesive subsoils, Broms [3.70] has prepared convenient nondimensional charts, reproduced as Fig. 3-24. From these charts, the ultimate lateral resistance for a given pier design may be determined in terms of the nondimen-

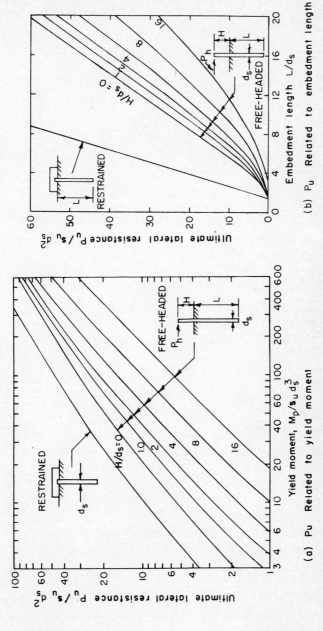

Fig. 3-24 Ultimate lateral resistance of cohesive soils. [*After Broms (1964).*]

(a) P_u Related to yield moment

(b) P_u Related to embedment length

sional parameter $P_u/s_u d_s^2$ related to both embedment length and the maximum flexural resistance of the pier. In instances where the embedment length controls the ultimate lateral resistance, the maximum bending moment may be established for the free- and fixed-head cases by Eqs. (3-56) and (3-57), respectively.

$$M_{\max}^+ = d_s P_u \left[1.5 + \frac{0.55 P_u}{s_u d_s^2} + \frac{H}{d_s} \right] \tag{3-56}$$

$$M_{\max}^- = 4.5 s_u d_s (L^2 - 2.25 d_s^2) \tag{3-57}$$

(b) Limit Design—Cohesionless Soils Nondimensional charts to compute the utlimate lateral resistance of the fixed- and free-head piers have also been prepared by Broms for cohesionless soils. These charts, reproduced as Fig. 3-25, express P_u in terms of the parameter $P/k_p d_s^3 \gamma$, which is related to both the embedment length and the flexural resistance of the pier. The smallest P_u and the corresponding failure mechanism can therefore be readily determined from charts a and b in Fig. 3-25 for any pier design.

Where the ultimate lateral resistance of a pier is controlled by embedment length, it is desirable to check the maximum bending moment that has to be sustained and, if economically appropriate, to modify the flexural resistance of the pier section. For the free-head and fixed-head cases, the maximum bending moment is expressed by Eqs. (3-58) and (3-59), respectively.

$$M_{\max}^+ = P(H + f) - \frac{\gamma d_s k_p f^3}{2} \quad \text{FREE HEAD} \tag{3-58}$$

$$M_{\max}^- = 0.67 PL \quad \text{FIXED HEAD} \tag{3-59} \text{ F}$$

As introduced in Art. 3-4, $f = 0.82(P/\gamma d_s k_p)^{1/2}$ represents the depth below the surface to the point of maximum positive moment.

(c) Safety Factors Determination of the allowable lateral load of a pier or group of piers is based on selection of a suitable safety factor to be applied to the calculated ultimate load resistance. The many uncertainties involved in predicting both the magnitude of the design loads and the soil resistance are, unfortunately, reflected in the selection of a suitable safety factor. Particular difficulty is experienced in determination of maximum lateral live loads, as, for example, would be imposed by wind, water, ice, and earthquake forces or by dynamic impact, such as berthing ships. It has been suggested [3.70] that overload factors can be used which are scaled on the probability of the occurrence of the design load during the economic life of the structure. More conventionally, safety factors of 1.5 and 2 are reported to be frequently applied to the calculated dead and live load, respectively.

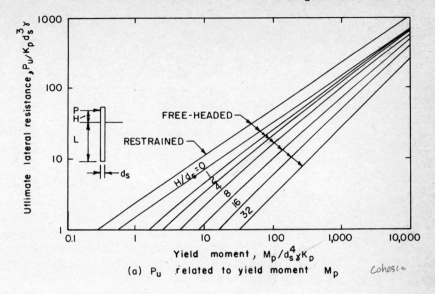

(a) P_u related to yield moment M_p Cohesic

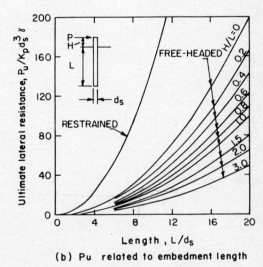

(b) P_u related to embedment length

Fig. 3-25 Ultimate lateral resistance of <u>cohesionless</u> soils. [*After Broms* (*1964*).]

As analysis of the ultimate lateral resistance is based on the soil resistance parameters which (particularly for cohesive soils) are dependent upon seasonal variations in water content, natural variations in properties, long-term creep, and sampling disturbance, it is good practice to reduce measured values. For cohesive soils, it is recommended that s_u determined from laboratory tests be reduced by one-third or one-fourth, the greater reduction applying to stiff fissured clays and to soft organic silts and clays. As the k_p factor for cohesionless soils is based on the measured or estimated friction angle ϕ, it is recommended that ϕ, as measured by laboratory tests, be based on values which are near residual strength rather than on peak strength values. Thus, a reduction in ϕ_{max} of 10 to 15 percent would be appropriate for dense to very dense sands.

3-5.5 Empirical Design Methods for Lateral Loads

There have been many empirical and semiempirical design techniques proposed for laterally loaded piles and poles which could be assumed to be equally applicable to drilled piers. Some of these techniques have been developed for relatively short and rigid members, such as would be used to resist loadings imposed by signboards. Most are based on ultimate load considerations. A few are derived from load-test results.

(a) **Outdoor Advertising Association of America** The best-known empirical design technique is that of the OAAA, derived from extensive testing of 10- to 18-in.-pipe poles with embedment depths of 4 to 7 ft. From the results of these tests, design charts were prepared which permit calculation of depth of embedment if the allowable soil reaction is known. The allowable soil stress is given in terms of the pullout resistance of a 1.5-in.-OD indicator auger or, less rigorously, is correlated with soil type as summarized by Osterberg [3.75].

Ivey and Hawkins [3.76], in describing the design of drilled piers for Texas Highway Department sign foundations, concluded that the OAAA design chart yields conservative results for short, rigid posts with diameters less than 24 in. (61 cm), embedments less than 10 ft (2.5 m), and horizontal loads less than 3 kips (1,360 kg). For design conditions not covered by the chart, they suggest that the OAAA derivation be used in equation form as

$$L = \frac{1.18P_h}{d_s S_1} + \sqrt{\left(\frac{1.18P_h}{d_s S_1}\right)^2 + \frac{2.63P_h H}{d_s S_1}} \tag{3-60}$$

$$\frac{S_1}{S_2} = \frac{0.28L}{H + 0.34L} + \frac{1}{2} \tag{3-61}$$

where L = depth of embedment
d_s = pier diameter
S_1 = average passive resistance in upper two-thirds of embedment
S_2 = average passive resistance in lower one-third of embedment
P_h = horizontal load
H = height of load above ground surface

In Eq. (3-61), S_1 is taken as two-thirds of the maximum stress in the upper two-thirds of the embedment, acting at $x = 0.34L$ and S_2 is assumed to be one-half the maximum stress on the bottom one-third of the embedment. The maximum stress reactions are compared to the passive soil resistance (such as would be calculated for a long wall) at depths of 0.34L, and L is adjusted until a factor of safety of at least 2 is obtained. This procedure results in a real factor of safety against soil failure of at least 5 for cohesionless soils and at least 10 for cohesive soils (see Art. 3-4).

(b) Building Code Criteria The Uniform Building Code specifies design equations which have a form similar to the OAAA criteria and are related to ultimate load considerations. Equation (3-62) for non-constrained (free-head) members and Eq. (3-63) for constrained (fixed-head) members are given, respectively, as

$$L = \frac{A}{2}\left[1 + \left(1 + \frac{4.36H}{A}\right)^{\frac{1}{2}}\right] \tag{3-62}$$

$$L^2 = \frac{4.35P_hH}{S_1d_s} \tag{3-63}$$

$S_2 = \text{BEAR @ } 1/3 \text{ depth of Imbed}$

The notations are as given above, with S_1 defined as the allowable lateral soil-bearing pressure at a depth equivalent to the depth of embedment and $A = 2.34P_h/S_1d_s$ (see Sec. 2905, Uniform Building Code, 1970).

3-5.6   Laterally Loaded Groups

Although, to the authors' knowledge, there is no reported experience concerning the lateral load behavior of full-scale pier groups, it is reasonable to assume a response somewhat similar to that reported for groups of piles.

Test data from laterally loaded piles indicate that the lateral resistance of a group at working loads can be calculated as the sum of the resistances of the individual units. The maximum bending moment of group piers, however, may be significantly greater than that calculated for a single pier using the soil modulus analysis. The increase in group pier moment is probably not significant at pier spacings greater than about three to four diameters, but increases rapidly with an increasingly smaller pier spacing. Group pier bending moment also increases with group size, as shown by Fig. 3-26.

Theoretical studies of the group action of rigid piers have been conducted by Prakash [3.74]. His findings indicate that group action will not develop if piers are spaced more than three diameters normal to the direction of loading and more than six to eight diameters parallel to the direction of loading.

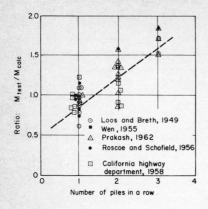

Fig. 3-26 Comparison of calculated and observed maximum moment in laterally loaded pile groups. [*After Broms (1964)*.]

3-6   DESIGN FOR TENSILE LOAD

The need to design piers to sustain tensile loads is commonly encountered in such engineering works as retaining walls, docking facilities, transmission line towers, and other elevated structures subject to horizontal loading. Resistance by straight-shaft piers to uplift is developed solely as shaft resistance and is dependent on many factors (see Art. 3-2), including shearing resistance of materials in contact with the pier, the amount of shaft movement, the properties of the pier, and the construction methods used. In practice, analysis of straight-shaft piers subject to tensile load is identical [3.44] to the methods used to analyze the shaft resistance of piers loaded in axial compression, although the results of one series of specially instrumented pile load tests [3.77] show the shaft load transfer to be about 30 percent greater during compressive loading. The influence of residual pile-driving stresses, however, may well explain the disparity in the tension as compared with compression load-test results.

Additional tensile load resistance is developed by the belled segment of an underreamed pier, but this requires an appreciably greater displacement than is necessary to mobilize the ultimate shaft resistance. If the pier depth is sufficient, the ultimate resistance of the materials in contact with the bell can be approximated from bearing capacity theory and, as in limit analysis for compressive loading, the total uplift capacity of a belled pier can be expressed as the sum of the shaft and bell resistance. A more commonly used alternative method expresses the ultimate tensile load resistance of a belled pier in terms of the frictional resistance and weight of a fictitious cylindrical pier of length L and diameter d_b. It is advisable to apply both methods to a design problem and to use the lowest resistance calculated by the two procedures.

The displacement compatibility method outlined in Art. 3-3 for compressive load can be applied to analysis of the uplift resistance of both straight-shaft and belled piers. However, in practice, the use of this method is inhibited by an almost total lack of documentation of the pier-to-soil load transfer during tensile loading. Elastic, elastoplastic, and finite element compatibility solutions are also theoretically applicable to various degrees, but each must depend on characterization of at least one soil or rock parameter which is nonlinear, stress-dependent, and influenced by pier-soil interaction. Because of the present state of development of displacement compatibility solutions, only the limit design approach is considered in the following paragraphs.

3-6.1   Straight-shaft Piers

Although stress conditions at the pier-soil (or -rock) interface under tensile loading would be expected to be somewhat different from those developed by compressive loading, Sowa [3.78] demonstrates, by a comparison of adhesion and undrained shear strength values, that there is a reasonable similarity, at least in short-term ultimate resistance (Fig. 3-27). Correspondingly, the ultimate tensile load of a straight-shaft pier in *cohesive* soils can be predicted from the expression

$$T_p = \pi d_b \Sigma \alpha s_u L + W_p \tag{3-64}$$

where T_p = ultimate tensile load and W_p = effective weight of pier. The results of this analysis, shown by Fig. 3-27, indicate good correlation with the lower limit of the relationship between the shear reduction

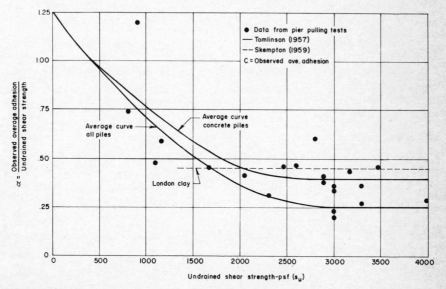

Fig. 3-27 Relationship between reduction factor α and undrained shear strength. [*After Sowa* (*1971*).]

factor (α) and undrained shear strength (s_u) proposed for compressive loading by Tomlinson [3.16].

The ultimate shaft resistance developed by a straight-shaft pier can be expressed for tensile loading in cohesionless soils as

$$T_p = \pi d_b \Sigma k_s \sigma'_v \tan \delta \, \Delta L + W_p \tag{3-65}$$

Consideration of the various soil and pier parameters contained within Eq. (3-64) is included in Sec. 3-3.1 for compression loading. Of these parameters, selection of the lateral earth pressure coefficient (k_s) is the most uncertain. There is some evidence that k_s is higher for piers with L/d_s less than 15 to 20, and that lower values for longer piers may be due to soil arching around the pier shaft. As in compressive loading, k_s would not be expected to exceed the coefficient of earth pressure at rest (k_0).

3-6.2   Underreamed Piers

In considering the uplift capacity of a belled pier as the sum of the ultimate shaft and bell resistance, the tensile bearing capacity (T_b) of the bell has been expressed for cohesive soils as

$$T_b = \frac{\pi}{4} (d_b^2 - d_s^2) N_c \omega s_u \tag{3-66}$$

and for cohesionless soils* as

$$T_b = \frac{\pi}{4} (d_b^2 - d_s^2) \sigma'_v N_q \tag{3-67}$$

provided the top of the bell is at least four to five bell diameters below the surface. The foregoing expressions are not theoretically rigorous and should be used with caution.

A shear strength reduction factor (ω) should be applied to account for disturbance of the bearing surface during underreaming. Until documented correlations are available, ω should not be more than $\frac{3}{4}$ and preferably not more than $\frac{1}{2}$. Because of a possible reduction in shaft resistance due to underreaming, it is also prudent to ignore development of shaft resistance in the portion of the shaft extending at least one bell diameter above the top of the bell.

The "friction cylinder" method of calculating ultimate uplift resistance conservatively assumes development of a vertically oriented cylindrical failure surface above the bell, with the diameter of the cylinder equal to the bell diameter. By this method, the ultimate tensile load capacity is expressed for cohesive soils as

$$T_p = \pi d_b L s_u + W_s + W_p \tag{3-68}$$

---

* Bells formed in uncemented cohesionless soils are uncommon and would require special drilling techniques.

and for cohesionless soils as

$$T_p = \pi d_b L k_s \sigma'_v \tan \phi + W_s + W_p \qquad (3\text{-}69)$$

where W_s = weight of soil within the cylinder and σ'_v is the effective vertical stress along the pier, assumed to reach a maximum at a depth of 15 to 20 shaft diameters below the surface. The effect of underreaming on the lower portions of the "friction cylinder" can be considered by a reduction in the shear strength (adhesion) of clay soils and a reduction in the coefficient of lateral earth pressure (k_s) and/or in the effective friction angle (ϕ') of cohesionless soils.

3-7 SPECIAL DESIGN CONSIDERATIONS

3-7.1 Swelling Soils and Rocks

The drilled pier is a favored form of foundation in areas of swelling soils and shales because it provides a relatively inexpensive way of taking structural loads down to stable material. However, there are other design considerations, in addition to depth to stable soil or rock, which may need to be taken into account. Some of them are:

(a) Uplift Resistance Where the pier is deep and the concrete of the shaft is cast in direct contact with an expansive soil, subsequent swelling of the soil can grip and lift the shaft, producing surprisingly high tensile stresses. The importance of this effect becomes evident when it is recalled that swelling pressures in clay soils sometimes reach 20 tsf (kgf/cm²) and more, and that free swell of such a soil may amount to 20 percent or more of the thickness of the zone of active heaving. Instances of unreinforced pier shafts broken in tension from this cause are common in many areas where swelling soils are prevalent. The break is often immediately above the base or underream.

The design of a drilled pier under these circumstances should include provision for uplift resistance in the base, either in the form of an enlarged base or in sidewall friction below the active zone, and sufficient reinforcement to resist any tensile forces which may be expected to develop.

(b) Isolation for Uplift Forces As a substitute for designing the pier to resist uplift forces, or as an extra safety measure to be used with such a design, special means for isolating the pier from contact with the heaving soil are used in many areas. One expedient that is often used is to drill the shaft hole oversize, use a cardboard or light metal liner as a form for the pier shaft, and fill the annular space around the liner with loose sand or pea gravel. This device probably works satisfactorily at first, but repeated seasonal cycles of soil heave and shrinkage could, conceivably, pack the filling to a point where it would grip the column tightly. And more important, the presence of a permeable layer around the shaft affords a free path for water—from storm, ponding, utility leakage, or other source— to reach the supposedly stable soil at the bottom of the pier and conceivably to heave the entire pier from the bottom.

Another device that seems more reliable is to use a light metal casing. Here the lower part of the drilled hole should be a force fit for the casing, which can be turned or pressed down. The upper part should be drilled or reamed larger, and the clearance space can be filled with something that will impede free infiltration of water—for example, a bentonite slurry or a heavy oil.

A design used with reported success in San Antonio, Texas, an area of very active clays, is illustrated in Fig. 3-28. A pipe or tubing section is introduced into the pier in lieu of reinforcing. The concrete-filled pipe is designed to carry the compressive load at the top. The outside of the pipe from the top down to the bottom of the expansive clay layer is coated with a bituminous mastic material. When the pier is gripped and lifted by the expansive clay, the annulus of concrete outside the coated section breaks in tension near the bottom of the expansive layer, the mastic coating shears or flows, and the upward force is limited by the shearing strength (or viscosity) of the mastic. Figure 3-29 illustrates application of the mastic coating. For data on an appropriate coating see p. 91.

The designer must remember that in an expansive soil area the "active" zone, even though it may not be thick enough to set up destructive tensile stresses by gripping the pier shaft, can still break a pier in tension by lifting an attached grade beam. Breaks thus produced are usually at the top of the pier or at the bottom of the dowels or top reinforcing. Because the forces involved can be very large, the only economical way of avoiding this kind of breakage is to isolate the grade beams from contact with the ground. This can be accomplished by casting the grade beams on top of hollow cardboard forms, which will crush without damage to the structure when the supporting soil expands. These special forms are generally available to builders in expansive clay areas.

(c) Concentration of Load to Balance Uplift In many instances it is possible to resist uplift forces developed in expansive clays or rocks by proportioning bearing surfaces to keep the imposed load *greater than* the expansive forces that may be developed. Because the possible expansive forces are affected by many variables (see p. 90), testing to directly determine expected expansive forces is rarely if ever practical. However, tests for a general determination of expansive potential

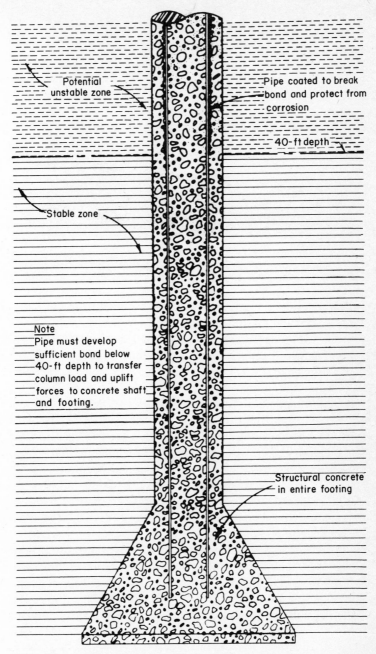

Potential
unstable zone

Pipe coated to break
bond and protect from
corrosion

40-ft depth

Stable zone

Note
Pipe must develop
sufficient bond below
40-ft depth to transfer
column load and uplift
forces to concrete shaft
and footing.

Structural concrete
in entire footing

Fig. 3-28 Design of belled pier for relief of uplift due to expansion of upper clay layer. Note that the outer annulus of concrete is expected to break in tension near the bottom of the expansive clay layer. (*Raba and Associates, Consulting Engineers, San Antonio, Tex.*)

Fig. 3-29 Bituminous (shearable) coating being applied to steel tube for interior of pier design shown in Fig. 3-28. (*Raba and Associates, Consulting Engineers, San Antonio, Tex.*)

are available [3.79], and often regional experience provides a satisfactory guide (see tabulation for Denver and eastern Colorado, Appendix E).

In proportioning bearing surfaces to resist uplift due to expansion, the designer must remember that only *real dead loads* will be effective; live loads cannot be counted on to exist during the period when expansive forces are in effect.

(d) Identification of Swelling Potential Swelling tendencies in soil and rocks are not only dependent on the nature and properties of the materials, but are also affected by such things as seasonal and climatic weather effects and cycles; groundwater regimen as affected by construction activities; stress relief as produced by excavations; chemical changes (especially in, but not restricted to, fills); freezing and thawing; buildup of mineral crystals; and capillary movement of moisture caused by covering an area by pavement, exposing an area to drying, or changing vegetation cover.

Because of the variety of these conditions, the possibility of swelling or heave of foundation soil should be determined by a geotechnical engineer or an engineering geologist who not only is familiar with the foundation and fill materials of the project area, but is also made aware of all relevant design and construction conditions contemplated for the project.

3-7.2 Downdrag Loads ("Negative Skin Friction")

Whenever drilled piers (or piles) pass through a consolidating soil formation, they can be subject to "downdrag" loads as the settling soils grip and hang up on the shafts. One circumstance in which this effect is common is in new construction through fill placed over a compressible formation; another is encountered when soil settlement is produced by drawdown of a water table and existing foundations have some of the weight of the settling soil transferred to them by sidewall shear.

In the case of new piers constructed through a settling formation, the magnitude of the possible downdrag load on single piers can be estimated by the same methods as those used to estimate shaft resistance under compressive loading (Art. 3-3). More important, development of these unwanted loads can be prevented by double-casing the pier shaft through the settling formation, or sometimes by the use of a viscous mastic coating on the casing. Bjerrum et al. [3.80] report experimental results for steel piles coated with a layer of about 1 mm of bitumen (described as "a straight run bitumen with the penetration 80/100"), in which the bitumen coating reduced downdrag forces ("negative skin friction") from settling fill to about 10 percent of the downdrag forces exerted on untreated piles. Unlike the situation with swelling soil formations, here the free passage for water provided by the space between two casings will do no harm.

3-7.3 Dynamic and Earthquake Forces

The analysis of dynamic and especially of earthquake forces on deep foundations—either drilled piers or piles—is very complex. Application of such analysis to design is, except in the simplest cases, a specialty which should be handled by an engineer accustomed to dealing with soil and rock dynamics problems. We cite these problems only in very general terms, with the objective of showing the reader what problems arise and giving him a general idea of the current approach to analysis of the forces involved. Nair [3.81] has presented a comprehensive review of the subject, with a list of 66 references to relevant technical papers and articles.

Of most significance in consideration of dynamic effects are the resulting horizontal stresses and motions. Vertical components of such forces have, in most cases, little effect on deep foundation design.

Dynamic forces that sometimes have to be considered in foundation design include:

(a) Machine Vibrations These are generally small-amplitude vibrations, which become important when the operating frequency approaches the natural resonance frequency of the subsoil and foundation system. This is usually a problem in soil dynamics, and the remedies do not usually involve pier design. The reader is referred to *Vibrations of Soils and Foundations* by Richart, Hall, and Woods, Prentice-Hall, Inc., Englewood Cliffs, New Jersey (1970).

(b) Wave and Wind Forces Here the forces involved definitely do affect pier design. The reader is referred to *Civil Engineering in the Oceans,* the Proceedings of an ASCE Conference in San Francisco, September, 1967, the sections "Waves and Response" and "Foundations."

(c) Earthquakes These are events in which horizontal displacements predominate, and vertical effects on drilled pier design are usually neglected, except for provision to resist vertical accelerations of structure with respect to pier. Designs are usually intended to provide for safe foundation support for assumed earthquake intensities. Protection against displacements occurring *in a fault zone* is not usually attempted. In earthquakes, the distorting forces are imposed through the ground and transmitted by foundation to structure.

In addition to the dynamic forces ordinarily imposed by earthquake distortions, another possible design consideration is the liquefaction of granular subsoils. This has occurred in a number of cases. This possibility can be predicted from preliminary soil investigations, and its consideration should never be overlooked in the design of foundations in earthquake-prone areas. The forces involved, if foundation liquefaction occurs, are in the form of downdrag and lateral forces on piers and loss of frictional support. They can usually be handled by static design techniques, but in extreme cases it might not be possible to handle them at all, and the risk of complete failure would have to be assessed and accepted—or the project be reevaluated.

(d) Design for Earthquake Effects The special requirements of drilled pier design (or any deep-foundation design) for regions of high seismic activity are difficult to determine on any rational basis. The common method of making an "earthquake allowance" by assuming a static horizontal load as a percentage (for example, 10 percent) of the vertical static load has no rational basis at all, but it is valuable nonetheless because it does require some strengthening of piers against earthquake deformations in earthquake zones. For important structures supported on long piers, preservation of vertical support in the event of a major earthquake becomes a major design objective and warrants the special effort and expense involved in applying the best available analytic techniques to determination of probable forces and deflections.

Analysis of soil response to earthquake movements, and of structural movements and stresses produced by the interaction of ground movements and structure inertia, is a technical specialty beyond the scope of this text. The recent developments of digital computer techniques, discrete element analysis, and dynamic soil testing—plus the accumulation of strong-motion earthquake records—provide the mathematical tools for these analyses. The general assumptions that are made for the purposes of an analysis for long piers through a deep weak soil layer are as follows:

1. Soil motion would be complex, consisting of higher frequencies and smaller amplitudes in the vertical direction, superimposed on larger amplitudes and lower frequencies in the horizontal plane; and the motion of most concern would be the horizontal.

2. Long piers (or piles) are assumed to deform horizontally with the soil—i.e., they will not be able to "cut through" the moving soil mass.

3. A "design earthquake" is assumed, represented by a horizontal acceleration record for a recorded earthquake in ground conditions as similar as possible to the subject site and of a magnitude believed to be appropriate to the seismic situation of the site (i.e., regional seismic activity, proximity to a major fault, etc.).

Figure 3-30 shows part of a strong-motion record that was selected as a design earthquake for the analysis of pile foundations for a building complex in Burlingame, California.

To analyze the effect of the design earthquake on a long pier (or pile) which is assumed to deform with the soil in which it is embedded, the soil profile is subdivided into a series of layers (25 in the example shown in Fig. 3-31). Each layer is represented by a mathematical analogue in the form of appropriate mass, stiffness, and damping parameters; and the system is then analyzed, using a digital computer, as a multi-

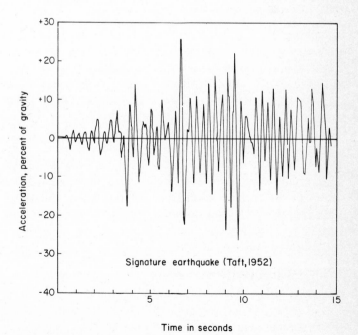

Fig. 3-30 Example of a design earthquake record.

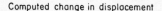

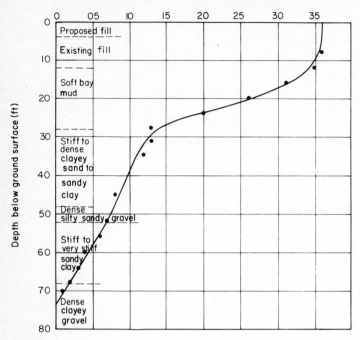

Fig. 3-31 Deflection curve for pier embedded in the soil profile (left-hand column) and subjected to the design earthquake illustrated in Fig. 3-30.

degree-of-freedom lumped-mass system. The computer output gives the time-history of motion for each of the layers.

If the maximum computed change in displacement from the input motion is then plotted for each layer and the computed points are joined by a smooth curve (Fig. 3-31), the resulting curve will be an approximation for the first-mode deflected shape of a long pier embedded in, and conforming to the distortion of, the soil profile. From this curve the pier diameter can be computed, along with the reinforcing required to prevent breakup of the shaft (with application of an appropriate factor of safety). A comparison of the soil profile and the deflection curve of Fig. 3-31 shows clearly how the stiffer layers suffer the smaller strains, and the highest bending moments are developed at the transitions between soft and stiff materials.

3-7.4   Corrosion Considerations

Wherever piers have to be founded in, or pass through, potentially corrosive soil or groundwater, special attention must be given to protection of the concrete against corrosion deterioration. Potentially troublesome materials include organic or highly acid soils; chemical wastes; and sulfate-bearing, especially gypsum-bearing, soils.

Many shales are potentially troublesome, including those containing pyrites and gypsum. Where such soil or rock conditions are known or suspected, the soil investigation should include tests for pH, resistivity, and sulfate content. Where local experience or the results of a foundation investigation indicate the possibility of contact of foundation elements with potentially corrosive soils, groundwater, or gases (H_2S, for example), the designer should be aware of the potential for trouble and should specify the use of ASTM Type II cement in foundation concrete that is to be exposed to sulfate solutions of 150 to 1,000 ppm, and ASTM Type V when the SO_4 concentration can exceed 1,000 ppm. It has been reported that the use of air entrainment improves the sulfate resistance of concrete made with any type of cement [3.82].

3-8 CONSTRUCTION CONSIDERATIONS FOR THE DESIGNER

There are a number of practical considerations relating to construction and inspection problems and techniques which the designer of drilled piers should have in mind. Most of them are details which, handled one way, can give trouble during construction and result in increased costs, and, handled another way, will help produce an economical, smooth-running job. Some of these are offered here as a checklist for designer and specification writer:

3-8.1 The Subsurface Investigation

The soil and foundation report should give special attention to the problems that will or might be encountered in construction due to soil, rock, or groundwater conditions (see Chap. 6). The designer should study this report, and the report as well as the boring records should be available to prospective bidders and to the successful contractor.

3-8.2 Contractors' Equipment and Capabilities

The designer should know the type of equipment and range of equipment size available in the area of the proposed structure and should be careful not to specify pier types or sizes or underream shapes that are not readily available. For large or important jobs, it may be economical to bring equipment from a distant city; but for most structures, piers that can be produced by locally available equipment and techniques are generally most economical.

Table 3-9 gives one drilling contractor's tabulation of hole sizes, bell sizes, and estimated unit bearing capacities that can be reached as a

TABLE 3-9   Maximum Economical Rig Capabilities*

Rig model, Calweld	Torque, lb-ft	Max shaft dia, ft	Max bell dia, ft	Unit bearing, TSF
75CH............	60,000	4	10.5	0–10
		4	8.5	10–15
		3	6.5	15–20
150CH..........	105,000	6	12	0–12
		5	10	12–20
		4	8	20–30
		3	6.5	30–40
175CH..........	170,000	8	15	12–20
		7	12	20–30
		6	10	30–40
		5	9	40–50
		4	8	50–?
185CH..........	280,000	9	14	30–40
		7	12	40–50
		5	10	50–?
200CH..........	400,000	15	?	to 120+

* Caisson Corporation, Pennsauken, N.J.

function of rig size and torque. (The torque values do not agree exactly with those in Appendix A, but the table will serve to show that the higher the torque rating, the larger the hole and the harder the rock that can be drilled with the machine.)

3-8.3   Hole and Underream (Bell)

(a) **Minimum Hole Sizes** In those instances where a man must enter the pier hole, it must be large enough to permit him to be lowered and raised without discomfort or danger, as well as to perform his inspection duties while he is in the hole. If he is to examine the bottom of the hole, he must be able to stoop or squat and touch the bottom with his hands. Thirty inches (76 cm) is about the minimum hole diameter that will permit this, although a 24-in. (61-cm) or 27-in. (69-cm) shaft diameter will permit examination of the sidewalls of a straight shaft, to within about 2 ft of the bottom, and of the bottom of an underreamed (belled) base.

(b) **Location and Plumbness Tolerances** The smaller the design tolerances for location and plumbness, the more trouble the contractor is going to have in meeting these specification details and the more trouble the inspector is going to have in checking them. These tolerances should be as liberal as practicable. Whenever it is found during construction that tight tolerances on these items are slowing operations, or are making it necessary to drill additional holes or redesign pier caps, the tolerances should be reviewed with the designer and liberalized if possible. The contractor may find it economical, and the designer acceptable, to enlarge the size of a crooked or out-of-plumb hole, or to add reinforcing, rather than to bridge the unacceptable pier with a beam and two additional piers. Sometimes an unacceptable hole can be corrected by pulling the casing, backfilling, and redrilling the hole.

Commonly specified tolerances are: for plumbness, 1 percent or 1.5 percent of length; for straightness, the same; for location of top, ½ in. for each foot of diameter [or sometimes a flat 3 in. (76 mm) regardless of diameter].

(c) Underream (Bell) Size and Shapes Machine-cut underreams are generally limited to three times the shaft diameter. This is a built-in limitation, inherent in the design of most "belling buckets." In the case of the "single-door" 60° underreamer, the underream diameter may be limited to twice the shaft diameter. Whenever an underream more than three times shaft diameter is specified, it requires either hand belling operations or the drilling of an oversize shaft. Both of these are expensive operations. Whenever an oversize underream is considered, consideration should also be given to the possibility of using instead a deeper straight-shaft or rock-socket pier, taking advantage of sidewall shear support.

As mentioned elsewhere, the authors believe that any of the commercially available bell shapes is satisfactory if properly installed. Limitation of the bell to one shape or style may be uneconomical if some of the prospective bidders are equipped only with tools for another style. Many building codes specify that a bell shall have a surface making an angle of not less than 60° with the horizontal. This is, in the authors' opinion, unnecessarily restrictive. The same codes usually specify that the edge thickness of the bell shall be not less than 12 in., another unnecessarily restrictive requirement and one that does not coincide either with usual available underreaming tool design (except for dome-shaped bells) or with common practice. The standards of the Association of Drilled Shaft Contractors, Inc., (Appendix D) show a toe height of 3 in. (7.6 cm) for shaft sizes 18 in. (46 cm) to 42 in. (106 cm), and 6 in. (15 cm) for shaft sizes 48 in. (122 cm) to 96 in. (244 cm).

(d) Avoidance of Multiplicity of Shaft and Underream Sizes and Shapes Occasionally a set of plans will show a multiplicity of shaft or underream sizes, usually as a result of the designer's attempting to keep a uniform footing pressure on the soil, or to equalize settlements by varying footing diameters and pressures, or else to save concrete by using the minimum possible diameter for each shaft. The result, however, is often to increase job cost rather than to save money. In spite of calling for more concrete, a job schedule showing only three or four shaft sizes and a limited number of underream diámeters will produce lower bids than will a schedule calling for twice as many sizes. Every time the pier contractor has to change shaft size he has to change the drilling tool and casing size. His work is delayed, he has to have more money tied up in equipment, and the job site is cluttered with casing and tools not in use.

3-8.4 Permanent Casing and Liners

(a) Provisions for Casing Left in Place The plans may require that casing be left in place for several reasons: it may be required by building code (Chicago, for example); it may be required by the owner's standard (General Services Administration, United States, for example); or it may be dictated by the presence of caving soils or artesian groundwater conditions, in combination with high column loads and a need to keep risk to a minimum. However, casing is an expensive item, and in some localities experience indicates that even temporary casing is not needed. In any event, the designer would do well to omit permanent casing as a mandatory item, except where required by building code or where the geotechnical engineer or the supervising engineer directs that it be left in place.

(b) Casing Sizes and Wall Thickness Casing for drilled pier shafts is commercially available welded steel pipe ("line pipe"), which is produced in quantity for the pipeline industry. This pipe is specified by *outside diameter*. When the casing is to be left in place as part of the pier, the *outside* diameter of the *casing* is the nominal shaft diameter. Designers should keep this in mind because casing specified with an inside diameter the same as the nominal shaft size would have to be specially fabricated and would cost two to four times as much as standard pipe. Standard wall thicknesses for available casing diameters are given in Appendix D. Many building codes specify that permanent casing wall thickness shall be not less than $\frac{5}{16}$ in. (7.9 mm).

(c) Casing in Lieu of Reinforcing—Allowable Stresses Building codes—and good practice—require casing steel to have a yield stress of not less than 33,000 psf (16.2 kgf/cm²) and require an allowance of $\frac{1}{16}$ in. (1.6 mm) for loss of wall thickness by corrosion where soil or groundwater conditions indicate corrosion potential. Allowable steel stress is computed using this reduced cross section.

3-8.5 Concreting

(a) Strength and Allowable Stress Compressive strength (28-day strength, or f'_c) of concrete for drilled piers is rarely specified at less than

3,000 psf (1.5 kgf/cm²) in today's practice, and much higher strengths are available at a price premium. When loads are high and drilling conditions difficult, the designer will sometimes find it economical to use a higher strength concrete—say, 4,000 or 5,000 psf (1.96 or 2.45 kgf/cm²) —and larger shafts, rather than a smaller shaft with reinforcing or permanent casing.

In many city building codes, the allowable design stress in concrete in drilled piers is limited to a value ranging from 22.5 percent to 33 percent of the 28-day compressive strength (f_c'), to compensate for placement effects which may weaken the concrete. However, when concrete is placed in a permanent steel casing, placement conditions have few uncertainties, and some codes permit an increase in concrete stress to as high as 0.40, or even 0.45, f_c'. This allowance applies only to concrete placed in permanent casing, not to concrete placed in thin metal liners.

(b) Mix Design Considerations The mix design for concrete used in drilled piers may need to be quite different from that specified for structural concrete on the same project. If there is a reinforcing cage, or if there is a large underream, the ability of the concrete to flow between reinforcing bars, or to completely fill the bell, is a matter of prime importance. For these piers, the concrete should have a slump of about 6 in. (15 cm), and aggregate size must not be large enough to impede flow between reinforcing bars. A maximum aggregate size of ¾ in. (2 cm) is appropriate under these circumstances.

In some cases, it is satisfactory to obtain the high slump needed, without loss of strength, by using a richer mix, but this ordinarily results in shrinkage on setting. If the pier is to carry its load by end-bearing only, this may not be objectionable; and in instances where the sidewalls of holes through rock are grooved, or are naturally very rough, a little shrinkage may do no harm. But, in general, it is better to avoid shrinkage by the use of appropriate additives, and in rock-socket piers where sidewall shear is depended on to carry load it may be desirable to use an expanding agent to obtain maximum lateral contact pressure. Although the authors believe that this may sometimes be appropriate, they do not know of any instance of its having been done.

For piers without reinforcing, or with no more than a short cage at the top, concrete with 4-in. slump is appropriate unless the size and shape of the underream are such that there is doubt of the ability of the concrete to flow and completely fill the bell. Although free fall of the concrete in placement is approved by ACI, PCA, and various governmental construction agencies—and segregation due to free-fall impact is not usually considered a problem—a special segregation-resistant design, such as the use of a low-slump gap-graded mix having

70 percent coarse aggregate, all *above* ¾ in. (2 cm) size, and 30 percent fine aggregate, has been recommended [4.17]. *But it must be remembered that such a mix is entirely inappropriate wherever there is any condition that might limit the flow of the concrete into all the space that it is expected to fill.*

3-8.6   Reinforcing

Under many circumstances, full-length reinforcing for compressive load in a drilled pier shaft serves no useful function, interferes with concrete placement, and adds substantially to foundation costs. The designer should include only as much reinforcing as is really needed and should minimize this by the use of high-strength concrete and increased shaft size. Where soils are firm and loads are entirely in compression, dowels for connecting piers to column bases may be the only steel that is required.

The state of California uses a 12-ft-long (3.6 m) reinforcing cage at the top of the pier and usually no reinforcing below that. This practice is also common in building design in California. There is a construction advantage to the use of this short cage, in that it can be placed after most or all of the concrete has been placed and often after the casing has been pulled. As is reported in Chap. 4, trouble with long reinforcing cages is a very common construction fault; sometimes they tend to rise when the casing is pulled, sometimes to "squat" and disappear when the casing is withdrawn.

If reinforcing is unavoidable, consideration should be given to the probable load distribution along the shaft, particularly in long piers. As shown by Fig. 3-1, the amount of load reaching the lower part of the shaft may be only a small percentage of the load applied to the pier, and the reinforcing can be proportioned accordingly.

If the amount of reinforcing needed is relatively large, consideration should be given to the use of rolled sections ("core steel") or permanent casing in lieu of reinforcing bars. Although the initial cost of casing or core steel for reinforcing is usually higher than that of the equivalent reinforcing bars, the costs associated with concrete and reinforcing cage placement difficulties in long piers may far outweigh the saving in initial cost. The use of permanent casing merits special consideration because of the higher allowable compressive strength in the concrete (p. 99).

In a pier designed for tensile loads, reinforcing cages may have to extend the full length of the pier.

The effects of concrete slump and aggregate size on the forces exerted on the cage during concrete placement have been discussed in the preceding section. The designer and the specification writer need to give some careful thought to the cage itself, and should bear in mind that the bursting pressure set up by a column of concrete that does not flow freely through the openings in the cage can approach 2.5 times the pressure of an equivalent column of water—and that, even when outflow of the concrete is relatively free, very substantial radial and vertical forces can be set up. Cages should be designed with these forces in mind (a consideration which would result in more welds, fewer ties, and heavier spiral reinforcing than commonly used), the design should not be left to the fabricator (usually a separate contract from the pier construction), and the inspector should be instructed to make sure the design is followed. The authors take the position that where reinforcing cages are needed they should be made and installed to proper specifications, and that where they are not needed they should be omitted from the plans and specifications.

REFERENCES

3.1 Whitaker, T., and R. W. Cooke, An Investigation of the Shaft and Base Resistances of Large Bored Piles in London Clay, *Symp. Large Bored Piles, London,* 1966.

3.2 Reese, L. C., and M. W. O'Neill, Field Tests of Bored Piles in Beaumont Clay, *ASCE Annual Meeting, Chicago,* Preprint 1008, 1969.

3.3 Vesic, A. S., Ultimate Loads and Settlements of Deep Foundations in Sand, *Proc. Symp. Bearing Capacity Settlement Found., Duke University, Durham, N.C.,* 1967.

3.4 Seed, H. B., and L. C. Reese, The Action of Soft Clay along Friction Piles, *Trans. ASCE,* vol. 22, 1957.

3.5 D'Appolonia, E., and J. P. Romualdi, Load Transfer in End-bearing Steel H-piles, *Proc. ASCE,* vol. 89, no. SM-2, 1963.

3.6 DuBose, L. A., Load Studies of Drilled Shafts, *Proc. Highway Res. Board,* 1955.

3.7 Mattes, N. S., and H. G. Poulos, Settlement of Single Compressible Piles, *Proc. ASCE,* vol. 95, no. SM-1, 1969.

3.8 Poulos, H. G., and N. S. Mattes, Behavior of Axially Loaded End-bearing Piles, *Geotechnique,* vol. 19, no. 2, 1969.

3.9 Poulos, H. G., and E. H. Davis, The Settlement Behavior of Single Axially-loaded Piles and Piers, *Geotechnique,* vol. 18, 1968.

3.10 Johannessen, I. J., and L. Bjerrum, Measurement of the Compression of a Steel Pile to Rock Due to Settlement of the Surrounding Clay, *Proc. 6th Intern. Conf. SM & FE, Montreal,* vol. 2, University of Toronto Press, Toronto, 1965.

3.11 Chandler, R. J., The Shaft Friction of Piles in Cohesive Soils in Terms of Effective Stress, *Civil Eng. Public Works Rev.,* January 1968.

3.12 Meyerhof, G. G., The Ultimate Bearing Capacity of Foundations, *Geotechnique,* vol. 2, no. 4, 1951.

3.13 Terzaghi, K., "Theoretical Soil Mechanics," John Wiley & Sons, Inc., New York, 1943.

3.14 Berezantsev, V. G., V. S. Khristoforov, and V. N. Golubkov, Load Bearing Capacity and Deformation of Piled Foundations, *Proc. 5th Intern Conf. SM & FE, Paris,* vol. 2, p. 11, 1961.

3.15 DeMello, V. F. B., Foundation of Buildings in Clay, State of the Art Volume, *Proc. 7th Intern. Conf. SM & FE, Mexico City,* 1969.

3.16 Tomlinson, M. J., The Adhesion of Piles Driven in Clay Soils, *Proc. 4th Intern. Conf. SM & FE, London,* vol. 2, 1957.

3.17 Woodward, R. J., R. Lundgren, and J. D. Boitano, Pile Loading Tests in Stiff Clays, *Proc. 5th Intern. Conf. SM & FE, Paris,* 1961.

3.18 Chuang, T. W., and L. C. Reese, Studies of Shearing Resistance between Cement Mortar and Soil, *Res. Rept.* 89-3, Center for Highway Research, University of Texas, 1969.

3.19 Burland, J. B., F. G. Butler, and P. Dunican, The Behavior and Design of Large Diameter Bored Piles in Stiff Clay, *Symp. Large Bored Piles, London,* 1966.

3.20 Kerisel, J., Deep Foundations—Basic Experimental Facts, *Proc. N. Am. Conf. Deep Found., Mexico City,* 1964.

3.21 Vesic, A. S., Load Transfer, Lateral Loads and Group Action of Deep Foundations, *Performance Deep Found.,* ASTM STP 444, 1968.

3.22 Meyerhof, G. G., Penetration Tests and Bearing Capacity of Cohesionless Soils, *Proc. ASCE,* vol. 82, no. SM-1, 1956.

3.23 Vesic, A. S., Tests on Instrumented Piles, Ogeechee River Site, *Proc. ASCE,* vol. 96, no. SM-2, 1970.

3.24 Matich, M. A. J., and P. K. Kozicki, Some Load Tests on Drilled Cast-in-place Concrete Caissons, *Can. Geotech. J.,* vol. 4, 1967.

3.25 Mohan, D., and G. S. Jain, Bearing Capacity of Bored Piles in Expansive Clays, *Proc. 5th Intern. Conf. SM & FE, Paris,* vol. 2, 1961.

3.26 Skempton, A. W., The Bearing Capacity of Clays, *Proc. Brit. Bldg. Res. Congr.,* vol. 1, 1951.

3.27 Schmertmann, J. H., Static Cone to Compute Static Settlement over Sand, *Proc. ASCE,* vol. 96, no. SM-3, 1970.

3.28 Chang, C. Y., and J. M. Duncan, Analysis of Soil Movement Around a Deep Excavation, *Proc. ASCE,* vol. 96, no. SM-5, 1970.

3.29 Janbu, N., Soil Compressibility as Determined by Odometer and Triaxial Tests, *Proc. European Conf. SM & FE, Weisbaden,* vol. 1, 1963.

3.30 Kulhawy, F. H., T. M. Duncan, and H. B. Seed, Finite Element Analysis of Stresses and Movements in Embankments during Construction, *Rept.* no. TE-69-4, University of California at Berkeley, Office of Research Services, 1969.

3.31 Bjerrum, L., Observed vs. Computed Settlements of Structures on Clay and Sand, unpublished lectures, M.I.T., 1964.

3.32 Ladd, C. C., Settlement Analysis for Cohesive Soils, *SM Division, Pub.* 270, M.I.T., 1964.

3.33 Deere, D. U., A. J. Hendron, F. D. Patton, and E. J. Cording, Design of Surface and Near-surface Construction in Rock, *8th Symp. Rock Mech., University of Minnesota,* 1967.

3.34 D'Appolonia, E., and A. G. Thurman, Computed Movement of Friction and End-bearing Piles Embedded in Uniform and Stratified Soils, *Proc. 6th Intern. Conf. SM & FE, Montreal,* vol. 2, 1965.

3.35 Salas, A. J., and J. A. Belzunce, "Resolution Theorique de la distribution des forces dans les pieux," *Proc. 6th Intern. Conf. SM & FE, Montreal,* vol. 3, 1965.

3.36 Nair, K., Load Settlement and Load Transfer Characteristics of a Friction Pile Subject to a Vertical Load, *3rd Panam. Conf. SM & FE, Venezuela,* vol. 1, 1967.

3.37 Coyle, H. M., and L. C. Reese, Load Transfer of Axially-loaded Piles and Clay, *Proc. ASCE,* vol. 92, no. SM-2, 1966.

3.38 Coyle, H. M., and I. H. Sulaiman, Skin Friction for Steel Piles in Sand, *Proc. ASCE,* vol. 93, no. SM-6, 1967.

3.39 Snow, R., Telltales, *Found. Facts,* vol. 1, no. 2, Raymond International, Inc., 1965.

3.40 Mansur, C. I., and A. H. Hunter, Pile Tests—Arkansas River Project, *Proc. ASCE,* vol. 96, no. SM-5, 1970.

3.41 Van Weele, A. F., A Method of Separating the Bearing Capacity of a Test Pile into Skin Friction and Point Resistance, *Proc. 4th Intern. Conf. SM & FE, London,* vol. 2, 1957.

3.42 Hanna, T. H., The Distribution of Load in Long Piles, *Soils,* vol. VI, no. 22-23, Paris, 1971.

3.43 Terzaghi, K., and R. B. Peck, "Soil Mechanics in Engineering Practice," John Wiley & Sons, Inc., New York, 1967.

3.44 Tomlinson, M. J., "Foundation Design and Construction," John Wiley & Sons, Inc., New York, 1969.

3.45 Skempton, A. W., A. A. Yassin, and R. E. Gibson, "Theorie de la force portante des pieux dans le sable," *Ann. Inst. Tech. Batiment,* Trav. Publ. vol. 6, 1953.

3.46 Poulos, H. G., Analysis of the Settlement of Pile Groups, *Geotechnique,* vol. 18, no. 4, 1968.

3.47 Hrennikoff, A., Analysis of Pile Foundations with Batter Piles, *Trans. ASCE,* vol. 115, 1950.

3.48 Bowles, J. E., "Foundation Analysis and Design," McGraw-Hill Book Company, 1969.

3.49 Grandholm, H., On Elastic Stability of Piles Surrounded by a Supporting Medium, *Ing. Vetenskaps Akad. Hand.* 89, Stockholm, 1929.

3.50 Gleser, S. M., Lateral Load Tests on Vertical Fixed Head and Free-head Piles, *Symp. Lateral Load Tests Piles,* ASTM STP 154, 1953.

3.51 Poulos, H. G., Behavior of Laterally Loaded Piles: I—Single Piles, *Proc. ASCE,* vol. 97, no. SM-5, 1971.

3.52 Hetenyi, Milclos, "Beams on Elastic Foundation," University of Michigan Press, Ann Arbor, 1946.

3.53 Reese, L. C., and H. Matlock, Non-dimensional Solutions for Laterally Loaded Piles with Soil Modulus Assumed Proportional to Depth, *Proc. 8th Tex. Conf. SM & FE, University of Texas,* 1956.

3.54 Reese, L. C., and A. S. Ginzbarg, Difference Equation Method for Laterally Loaded Piles with Abrupt Changes in Structural Rigidity, EPR, *Memorandum Rept.* 39, Shell Development Company, Houston, Texas, 1958.

3.55 Palmer, L. A., and J. B. Thomson, The Earth Pressure and Deflection Along the Embedded Length of Pile Subject to Lateral Thrust, *Proc. 2nd Intern. Conf. SM & FE, Rotterdam,* 1948.

3.56 Matlock, H., and L. C. Reese, Generalized Solutions for Laterally Loaded Piles, *ASCE Trans.* 127, pt. I, 1962.

3.57 Davisson, M. T., and H. L. Gill, Laterally Loaded Piles in a Layered Soil System, *Proc. ASCE,* vol. 86, no. SM-5, 1960.

3.58 Davisson, M. T., and S. Prakash, A Review of Soil-pole Behavior, *Highway Res. Record,* no. 39, 1963.

3.59 Davisson, M. T., and J. R. Salley, Lateral Load Tests on Drilled Piers, *Performance Deep Found.,* ASTM STP 444, 1968.

3.60   Roscoe, K. H., and R. K. Schofield, The Stability of Short Pier Foundations in Sand, *Brit. Welding J.*, August 1956.

3.61   Ivey, D. L., K. J. Koch, and C. F. Raba, Resistance of a Drilled Shaft Footing to Overturning Loads—Model Test and Correlation with Theory, *Res. Rept.* 105-2, Texas Transportation Institute, Texas A & M University, 1968.

3.62   Terzaghi, K., Evaluation of Coefficients of Subgrade Reaction, *Geotechnique,* vol. 5, 1955.

3.63   Broms, B. B., Lateral Resistance of Piers in Cohesive Soils, *Proc. ASCE,* vol. 90, no. SM-2, 1964.

3.64   Vesic, A. B., Bending of Beams Resting on Isotropic Elastic Solid, *Proc. ASCE*, vol. 87, no. EM-2, 1961.

3.65   Francis, A. J., Analysis of Pile Groups with Flexural Resistance, *Proc. ASCE,* vol. 90, no. SM-3, 1964.

3.66   Davis, E. H., and H. G. Poulos, The Use of Elastic Theory for Settlement Prediction Under Three-dimensional Conditions, *Geotechnique,* vol. 18, no. 1, 1968.

3.67   Alizadeh, M., Lateral Load Tests on Instrumented Timber Piles, *Performance Deep Found.,* ASTM STP 444, 1968.

3.68   Broms, B. B., Lateral Resistance of Piles in Cohesionless Soils, *Proc. ASCE,* vol. 90, no. SM-3, 1964.

3.69   McClelland, B., and J. Focht, Soil Modulus for Laterally Loaded Piles, *Proc. ASCE,* vol. 82, no. SM-4, 1956.

3.70   Broms, B. B., Design of Laterally Loaded Piles, *Proc. ASCE,* vol. 91, no. SM-3, 1965.

3.71   Hansen, Brinch, The Ultimate Resistance of Rigid Piles against Transversal Forces, *Danish Geotech. Inst. Bull.* 12, 1961.

3.72   Matlock, H., and L. C. Reese, Foundation Analysis of Off-shore Pile Supported Structures, *Proc. 5th Intern. Conf. SM & FE, Paris,* 1961.

3.73   Reese, L. C., and W. R. Cox, Soil Behavior from Analysis of Tests of Uninstrumented Piles Under Lateral Loading, ASTM STP 444, 1968.

3.74   Prakash, S., Behavior of Pile Groups Subject to Lateral Loads, Ph.D. thesis, University of Illinois, 1962.

3.75   Osterberg, J. O., Discussion, Supplement to Symposium on Lateral Load Tests on Piles, ASTM STP 154-A, 1954.

3.76   Ivey, D. L., and D. L. Hawkins, Signboard Footings to Resist Wind Loads, *Civil Eng.,* pp. 34–35, December, 1966.

3.77   Hunter, A. H., and M. T. Davisson, Measurement of Pile Load Transfer, *Performance Deep Found.,* ASTM STP 444, p. 106, 1968.

3.78   Sowa, V. A., Pulling Capacity of Concrete Cast In Situ Bored Piles, *Can. Geotech. J.,* 1970.

3.79   Krazynski, L. M., and L. J. Lee, Expansive Soil Testing, *The Importance of the Earth Sciences to the Public Works and Building Official,* Seminar of the Association of Engineering Geologists, Anaheim, Calif., pp. 100–171, Oct. 18–23, 1966.

3.80   Bjerrum, L., I. J. Johannessen, and O. Eide, Reduction of Negative Skin Friction on Steel Piles to Rock, *Proc. 7th Intern. Conf. SM & FE, Mexico City,* vol. 2, 1969.

3.81   Nair, K., Dynamic and Earthquake Forces on Deep Foundations, *Performance Deep Found.,* ASTM, STP 444, p. 229, 1968.

3.82   Woods, H., Durability of Concrete Construction, *ACI Monograph* no. 4, Iowa State University Press, 1968.

SYMBOLS USED IN CHAPTER 3

SYMBOL	NAME	DIMENSIONS
a	Area	Length^2
a_b	Area of base	Length^2
a_s	Area of shaft	Length^2
A_m	Moment coefficient for linearly increasing K, applied P_t	
A'_m	Moment coefficient for constant K, applied P_t	
A_s	Shear coefficient for linearly increasing K, applied P_t	
A_y	Deflection coefficient for linearly increasing K, applied M_t	
A'_y	Deflection coefficient for constant K, applied M_t	
B_g	Breadth of pier group (see also W_g)	Length
B_m	Moment coefficient for linearly increasing K, applied P_t	
B'_m	Moment coefficient for constant K, applied P_t	
B_s	Shear coefficient for linearly increasing K, applied P_t	
B_y	Deflection coefficient for linearly increasing K, applied M_t	
B'_y	Deflection coefficienct for constant K, applied M_t	
C_y	A deflection coefficient for various degrees of fixity	
d_b	Diameter of base	Length
d_s	Diameter of shaft	Length
D_r	Relative density of cohesionless soil	
E	Young's modulus	Force/length^2
E_b	Deformation modulus of material below base	Force/length^2
E_c	Deformation modulus of concrete	Force/length^2
E_f	Efficiency of group of piers (or piles)	
E_i	Initial tangent modulus	
E_r	Deformation modulus of rock *in situ*	Force/length^2
E_s	Deformation modulus of material around shaft	Force/length^2
E_{seis}	Rock modulus calculated from seismic velocity	Force/length^2
E_t	Tangent modulus	
f	Depth below ground to M_p	Length
f_c	Static cone resistance, frictional component	Force
f'_c	Compressive strength of concrete	Force/length^2
f_s	Unit shaft resistance	Force/length^2
F_b	Safety factor for base load	
F_s	Safety factor for shaft load	
g	Gravitational constant 32.2 ft/sec^2	Length/time^2
H	Height of lateral load above ground	Length
I	Moment of inertia of pier section	Length^4
I_{bb}	Base deformation influence factor for base load	
I_{bs}	Base deformation influence factor for shaft load	
I_p	Top deformation influence factor for total load	
k	Coefficient of subgrade reaction	Force/length^3
k_1	Coefficient of subgrade reaction from tests on 1-ft-square plate	Force/length^3
k_0	Coefficient of earth pressure at rest	
k_p	Pier stiffness factor	
k_p	Coefficient of passive earth pressure	
k_s	Coefficient of earth pressure normal to shaft	Force/length^3
K	Soil modulus	Force/length^2
K_e	Coefficient in modulus of soil deformation formula	Force/length^2
K_1	Soil modulus determined from k_1	Force/length^2
K_L	Soil modulus at base of pier	Force/length^2

SYMBOL	NAME	DIMENSIONS
K_x	Soil modulus of depth x	Force/length^2
L	Length of pier (or depth of embedment)	Length
L_g	Length of group of piers (or average depth embedment)	Length
L_0	Depth from surface to center of rotation	Length
L'	L, rigid pier, measured from $1.5d_s$ below ground	Length
L'_0	Depth to center of rotation measured as above	Length
m_x	Moment at depth x	Force $\cdot$ length
M_t	Moment applied at top of pier	Force $\cdot$ length
M_p	Maximum resisting moment	Force $\cdot$ length
M_p^+	Maximum positive resisting moment of the pier	Force $\cdot$ length
M_p^-	Maximum negative resisting moment of the pier	Force $\cdot$ length
M_{max}^+	Maximum positive bending moment imposed	Force $\cdot$ length
M_{max}^-	Maximum negative bending moment imposed	Force $\cdot$ length
n	An exponent in modulus formula	
n_g	Number of piles in group	
n_h	Coefficient of horizontal subgrade reaction	Force/length^3
N	Standard penetration resistance	
N_γ, N_c, N_q	Bearing capacity factors	
N_{qh}	Lateral bearing capacity factor	
p	Unit soil reaction	Force/length^2
p_a	Atmospheric pressure	Force/length^2
p_b	Unit load of base	Force/length^2
p_x	Unit soil reaction at depth x	Force/length^2
P	Axial compressive load on pier	Force
P_b	Load carried by base	Force
P_g	Working load for group of piers	Force
P_h	Applied horizontal load on pier	Force
P_s	Average compressive load in the shaft	Force
P_t	Applied horizontal ground line shearing force	Force
P_u	Ultimate lateral resistance of pier	Force
P_x	Soil reaction at depth x	Force
Q	Axial load	Force
q_c	Static cone resistance, point component	Force
q_h	Lateral bearing capacity	Force/length^2
R	Stiffness factor for constant soil modulus	Length
R_f	Failure ratio—hyperbolic stress-strain representation	
s	Slope of pier	
s_u	Undrained shear strength of clay	Force/length^2
S_1	Average passive resistance, top two-thirds of embedment	Force/length^2
S_2	Average passive resistance, bottom one-third of embedment	Force/length^2
T	Stiffness factor for linearly increasing soil modulus	Length
T_b	Tensile bearing capacity of bell	Force
T_p	Ultimate tensile load on straight-shaft pier	Force
v	Shear	Force
v_x	Shear at depth x	Force
V_p	Compressional wave velocity	Length/time
V_s	Shear wave velocity	Length/time
V_L	Sonic velocity, laboratory tests on rock specimens	Length/time
w'_n	Natural water content	
W_g	Width of group of piers (see also B_g)	Length

SYMBOL	NAME	DIMENSIONS
W_p	Effective weight of pier	Force
x	Depth	Length
y	Lateral deflection	Length
y_x	Lateral deflection at depth x	Length
y_0	Lateral deflection at top of pier	Length
Z	Depth coefficient x/T	
Z_{max}	Depth coefficient L/T	
Z'	Depth coefficient x/R	
Z'_{max}	Depth coefficient L/R	
α	Shear strength reduction factor for shaft support materials	
α_r	Shear reduction factor for shaft support in rock	
β	Seismic modulus reduction factor	
γ	Unit weight of soil	Force/length³
δ	Angle of friction between pier and soil	
Δ_{bb}	Displacement of pier base due to deformation of soil under base load	Length
Δ_{bs}	Displacement of pier base due to deformation of soil under shaft loads (sidewall shear strain)	Length
Δ_p	Displacement of pier top	Length
Δ_r	Elastic rebound of pier top	Length
Δ_s	Elastic shortening of shaft	Length
Δ_L	Increment of shaft length	Length
λ	Shaft load influence factor	
μ	Poisson's ratio	
σ_0	Overburden pressure	Force/length²
σ'_v	Effective vertical pressure	Force/length²
σ_1, σ_3	Principal stresses	Force/length²
τ	Shear resistance along shaft	Force/length²
ϕ	Angle of internal friction of soil	
ω	Shear strength reduction factor for base support materials	

CHAPTER FOUR

Construction Equipment and Techniques

4-1 EQUIPMENT

4-1.1 Drilling Rigs: Types, Ratings, Capacities

Commercially produced drilling rigs of sufficient size and capacity to drill pier holes come in a wide variety of mountings and driving arrangements. Mountings are usually truck, crane, crawler, or skid (or occasionally wheeled trailers)—see Figs. 4-1 to 4-6. Driving arrangements usually fall into one of three classes: the kelly driven by a mechanically geared rotary table (Figs. 4-1, 4-3, and 4-4), the kelly driven by a yoke turned by a ring-gear (Figs. 1-2, 4-5, and 4-6), or the hydraulic drive, with the hydraulic motor either mounted at the turntable (Fig. 4-2) or mounted on top of the drill stem and riding up and down with it (Fig. 4-6). Choice of these details is a matter of contractor's preference, based on experience and suitability of the machine to job and ground conditions in the areas where it will be used and on local availability. But the choice of size (or capacity) of the drilling machines to be used on a job may become the concern of the owner, the architect, and the structural engineer as well as that of the contractor. The authors have encountered many instances where the successful bidder on a drilled pier contract did not have suitable or heavy enough

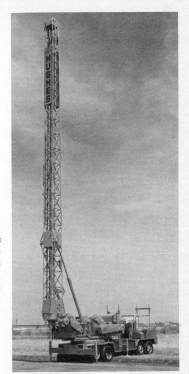

Fig. 4-1 Hughes (Williams) Digger LLDH-120T. This truck-mounted rig is rated as being capable of drilling shafts up to 10 ft (3 m) in diameter, to depths in excess of 100 ft (30 m). (*Hughes Tool Company, Inc., Houston, Tex.*)

Fig. 4-2 Texoma Model 600 with mast laid down for traveling. This rig has a mechanically geared rotary mechanism driven by a hydraulic motor, and a hydraulic cylinder for downward thrust on the kelly ("crowd"). (*Texoma, Inc., Sherman, Tex.*)

Fig. 4-3 Calweld Model 200-CH crane attachment. This machine has a torque output reported at 450,000 lb-ft (62,000 kgf-m). (*CALWELD, Division of Smith Industries International, Inc., Santa Fe Springs, Calif.*)

equipment for the job; and the result was invariably controversy, claims for "extras," or delayed completion—and very often all three.

Drilling machine ratings as presented in the manufacturers' catalogs and technical data sheets are usually expressed as "maximum hole diameter," "maximum depth," and "maximum torque" at some particular rpm (see Appendix A). The "maximum" data given are limiting values, usually dictated by dimensional, strength, and power limitations. The manufacturer's intent is not to recommend that the machine be used regularly for work at these ratings, but rather to warn the user never to exceed these limits and to approach them only occasionally and under favorable drilling conditions. A drilling machine working steadily near its upper limits of capacity in any category is inefficient and liable to mechanical failures. The curves in Fig. 4-7 show how drilling costs *per foot of hole* in a specific soil condition vary with hole diameter for a given machine, and how changing from the wrong size of drilling machine to one of more suitable capacity can reduce costs. (A similar diagram showing cost per cubic yard of soil removed would show lower minimum unit costs for successively larger machines.)

For auger drilling in rock, hardpan, or very hard soil, the available torque at auger cutting speeds becomes a major criterion of suitability. A contract involving hard drilling should never be let unless it is certain that the contractor has equipment that is sufficiently powerful to work within its economical range under job conditions. In the Philadelphia area, for example, it is common practice to drill 60-in.-diameter (1.5 m) holes through overburden and very dense decomposed mica gneiss/schist. The underlying unaltered rock is currently considered capable of supporting unit loads of 30 to 40 tsf (or kgf/cm²). In order to drill such holes, an auger machine with an available torque of 100,000 lb-ft (14,000 kgf/m) is required. Experience has shown that lighter rigs cannot drill holes of this size to the required depth in these formations.

The use of an auger with a drilling machine of suitable torque capacity also has proved much more economical than the use of core barrels or other tools with a lighter rig. Auger drilling machines with a torque rating of 360,000 lb-ft and higher are available in some cities (Fig. 4-3) [4.10, 4.11, 4.27].

The torque criterion mentioned above applies, of course, only to a machine using an auger-type tool; machines using core barrels, roller

Fig. 4-4 Watson crane-mounted Model 500CA, drilling an 8-ft (2.5 m) diameter hole in stiff clay soil. The auger is beginning to spin off a load of spoil; note the distance to which clods have been thrown during earlier trips. (*Watson Manufacturing Company, Fort Worth, Tex.*)

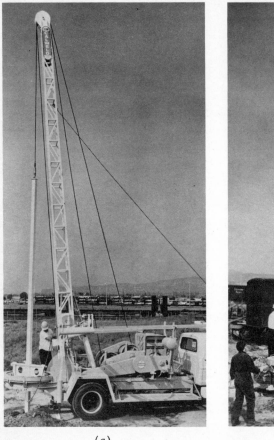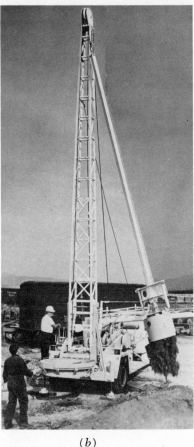

(a) (b)

Fig. 4-5 Calweld Model 150-B bucket auger rig, (a) drilling and (b) dumping its bucketload of spoil. Note the ring gear, through which the drill bucket is lifted for emptying. (*CALWELD, Division of Smith Industries International Inc., Santa Fe Springs, Calif.*)

bits, or down-hole chopping bits will not require as much torque—but will usually not make hole as fast as an auger *of suitable capacity* in hard soil or soft rock formations.

In Appendix A, auger drilling rigs from some of the major suppliers of this type of equipment in the United States have been tabulated, with some dimensional data and rated maximum capacities (manufacturers' data). These have been separated into heavy-duty, medium-duty, and light-duty, using the manufacturers' recommended maximum hole diameter and depth as criteria.

In addition to the auger-type drilling rigs, there are several models of rotary rigs using a circulating drilling fluid to carry the cuttings out of the hole which are used for pier drilling. Two of these, the

Acker Model WA and the Wabco 1500 HD, are illustrated in Figs. 4-6 and 4-8.

4-1.2   Rig Mountings

By far the largest number of pier-drilling machines in the United States are truck-mounted.   Their chief—and very important—advantage over other types is that of mobility: the ability to move at highway speeds between sites and, under favorable circumstances, to maneuver easily from hole to hole within a site.   Most of them are equipped also with sliding and rotating mounts, and the masts of many models can be tilted to drill battered holes—in some cases past the horizontal position. These ready adjustments make positioning of the auger over the hole location very fast and easy.   Figures 4-1, 4-2, 4-5, and 4-6 illustrate some of the truck-mounted rigs available.

Crane-mounted rigs must be brought to the site by heavy equipment trailers and are therefore less mobile and less adaptable to small jobs

Fig. 4-6   Acker Model WA hydraulic rig, drilling pier holes using multiple-roller bit, reverse circulation, and drilling mud.   (*Acker Drill Company, Inc., Scranton, Pa.*)

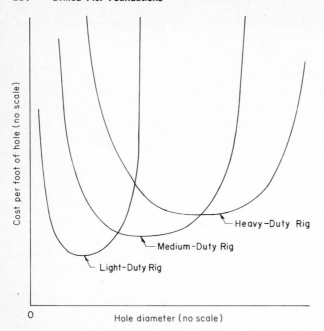

Fig. 4-7 Curves showing how cost per foot of hole varies
with capacity of drilling rig. Each size rig has limiting
minimum and maximum hole size, and the cost per foot of
hole increases rapidly as either limit is approached.

than truck-mounted machines. Most of these machines are furnished
as "crane attachments," to be mounted on a crane of the contractor's
selection.

Crane-mounted rigs are designed to handle large and heavy augers
and buckets, and in general can drill deeper holes and larger diameter
holes than rigs with other types of mounting. Their driving machinery
is generally higher off the ground than that of truck-mounted machines,
and consequently they can handle taller augers and underreamers. The
higher lifting capacity of the crane also allows handling of larger and
heavier casing sections and reinforcing cages without the necessity of
having a separate crane on the site for that purpose.

In some areas, most of the cranes available are themselves truck-
mounted, but more often they are carried on tracks and are therefore
readily maneuverable over soft ground.

Care is required in selecting a crane for a particular rig, or in adapting
a crane to mount a rig, to see that no part of the crane is overstressed
by application of forces that were not taken into account in the design
of the crane or its boom. This requires special attention to the stresses
that can be set up at the points of attachment of rig to crane. Some

crane-mounted rigs have the entire rig weight suspended from the crane; others are so mounted that their weight is carried near the base of the crane boom and on the cab.

Three large crane-mounted rigs are illustrated in Figs. 1-3, 1-4, 4-3, and 4-4.

Crawler-mounted rigs also have to be brought to a site by heavy equipment trailers and are therefore less mobile than truck-mounted equipment. Their on-site mobility is excellent, and because of the higher mounting, they can handle taller augers and underreaming tools than most truck-mounted rigs.

Specially mounted rigs for drilling pier holes in restricted spaces, areas with limited overhead, or on slopes are available in many cities. An example of these is the sidehill rig shown in Fig. 4-9.

4-1.3   Kellys and Kelly Extensions

Single-piece kellys are usually square driving shafts of solid steel, although hollow kellys are sometimes used for maximum torsion capacity for minimum weight. Single-piece kellys are made to drill to depths of 65 ft (20 m) or more. When deeper holes are needed, some manu-

Fig. 4-8 Wabco 1500 HD Holemaster drilling rig. This rig is used for drilling holes with roller-type bits, using drilling mud and reverse circulation, especially in the "limerock" areas of Florida. (*Wabco/Drilling Equipment Div., an American-Standard Company, Enid, Okla.*)

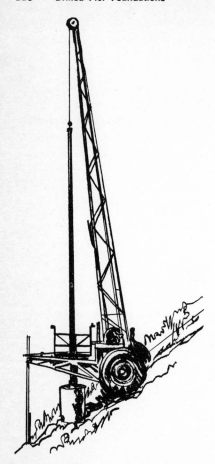

Fig. 4-9 Hillside drilling rig. (*Raby Hillside Drilling Inc., Los Angeles, Calif.*)

facturers use telescoping kellys, the inner square shaft sliding in a larger hollow square section. Other rig makers prefer to add pin-connected sections of drill shaft as needed. The Williams LLDH machine is equipped with a telescoping kelly that will drill to a depth of 120 ft (36½ m) without adding drill rods.

4-1.4   Downward Force; "Crowd"

The combined weight of the auger bit or bucket and the kelly usually provides sufficient downward force for good, productive rates of pier-hole drilling through normal soil formations. But when the rate of penetration is slowed by harder formations, it becomes necessary to add downward thrust to maintain bit penetration at an economical rate. This thrust—usually called "crowd"—is generally added by the use of one or more hydraulic cylinders reacting against the weight of the machine carrying the drilling equipment. The amount of "crowd" truck-mounted rigs can apply is limited by the force that is sufficient to pick up the rear end of the truck and its load. The tools, kelly, and drill stem of a crane-mounted rig are usually very heavy, and for many machines and in many applications, no additional "crowd" is required. Indeed, in most soil-drilling and many rock-drilling applications, the weight of bit, kelly, and drill stem (if any) may be more than is needed for good penetration; and often part of this weight has to be carried suspended from the crane and allowed to bear on the drilling surface only as needed. Some manufacturers augment this "tool" weight by the use of hydraulic cylinders reacting against the weight of the drilling machine and the crane. Others take care of the requirement by adding weight to the tools, declaring that suspended weight will drill a straighter, more perfectly vertical hole than will any "crowd" mechanism.

In the drilling of large-diameter pier holes, a major reduction in the weight of tools resting on the drilling surface may be required in order to avoid overloading the individual cutting teeth of a bit or core barrel. When a very large multiroller bit is being used in hard rock, however, a major addition to tooth pressure may be necessary. The former has been accomplished by building a flotation tank into the top of the drilling tool [4.26], and the latter by weighting the drilling tool with steel punchings or with lead [4.2, 4.3].

4-1.5 Drilling and Auxiliary Tools

Drilling Buckets The bucket-type drill, shown in Fig. 4-5, was used for almost all pier drilling in California for many years. The key to its success was the ingenious driving mechanism for turning the kelly: a ring-gear, large enough for the drilling bucket to be lifted out of the hole and up through the gear, so that it could be swung aside and emptied into a spoil pile or into a truck. This drive was the basis for the Calweld bucket auger machine (Figs. 1-2 and 4-5).

The efficiency of the "flight auger" (see below), which could be emptied by spinning it as it came out of the hole; the development of machines such as the front-end loader for picking up the spoils; and the demand for ever larger-diameter holes: all these trends have led to the development of machines which can be used with either bucket or auger. Drill buckets are still used in preference to flight augers under many circumstances, as they are often more efficient in soft soils or running sands, and the bucket is a standard tool for supplementing the flight auger in circumstances where the cuttings are too fluid or too loose to be brought out efficiently with an open helix.

Some drilling buckets are so designed that they can be used also for bailing water.

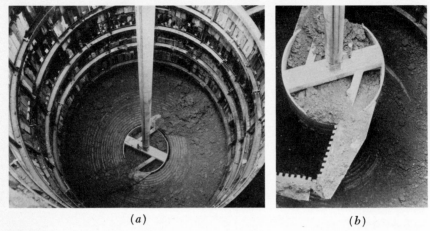

(a) *(b)*

Fig. 4-10 72-in. (183 cm) drill bucket, with hinged reamers for drilling 20.6-ft-(6.28 m) diameter access shafts. (*a*) Drilling with reamers. (*b*) Coming out with a bucketload of spoil. (*CALWELD, Division of Smith Industries International, Inc., Santa Fe Springs, Calif.*)

Hole Reamers for Drilling Buckets When large-diameter holes are drilled with machines using a ring-gear and yoke driving mechanism (Calweld, Earthdrill, etc.) and a drilling bucket (rather than an open-helix auger), the diameter of the drilling bucket is limited by the inside diameter of the ring-gear through which the bucket must be lifted. The hole diameter, however, can be enlarged by using a reaming tool which is attached to the bucket. The reamer cuts out the full diameter of the hole and brings the cuttings into the drill bucket, which is used to bring out and dump the spoils. Holes of very large diameter can be drilled in this manner (Fig. 4-10). Some manufacturers' catalogs list reamer diameters up to 17 ft (5.2 m).

Open-helix Augers ("Flight Augers") Many drilled pier shafts through soil or soft rock are now drilled with the open-helix auger. This tool may be equipped with a knife-blade cutting edge for use in most homogeneous soils, or with hard-surfaced teeth for cutting stiff or hard soils, stony soils, or soft to moderately hard rock. These augers are available in diameters up to 10 ft or more. Several commercially available models are shown in Fig. 4-11.

Underreamers Underreaming tools (or "buckets") are available in a variety of designs. Generally, there are three main shapes of footing which can be cut with commercially available tools: the dome-shaped footing, (approximately hemispherical, with a small cylindrical pilot depression at the bottom); the 30° underream (the sides making an angle of 30° with the vertical); and the 45° underream. The choice of which to use seems to be principally dependent on local custom and experience, and availability. Some contractors prefer the 45° tool, claiming that because it removes a smaller volume of soil for the same bottom-bearing area, it saves time in drilling and money in concreting. Others declare that the 30° tool (or the tool that cuts a dome) will drill faster and make a cleaner hole.

For large underreams, some contractors report that the bucket-type underreamer, which cuts a dome-shaped footing, carries out more soil per trip than the other designs and is therefore the fastest. This advantage has to be balanced against the cost of the additional concrete required by this shape of footing.

Some of the available tools cut a flat bottom, without any tendency to deepen the pier hole as they cut. Others cut a small sump at the bottom which is supposed to facilitate collection and removal of cuttings. Some underreaming tools have a tendency to deepen the hole or to pack cuttings under the bucket and create a false bottom of remolded and weakened material. Most underreamers cut a footing with a 6- or 12-in.-high (15 or 30 cm) cylindrical section at the bottom. Figure 4-12 shows two designs of underreamers.

Most underreaming tools are limited in size to a diameter three times the diameter of the shaft; and when larger bell diameters are required, either a larger shaft must be drilled or the machine-cut bell must be enlarged by hand labor (see Art. 4-2).

Core Barrels When rock becomes too hard to be removed with auger-type tools, it is often necessary to resort to the use of a core barrel. This tool is a simple cylindrical barrel, set with tungsten carbide

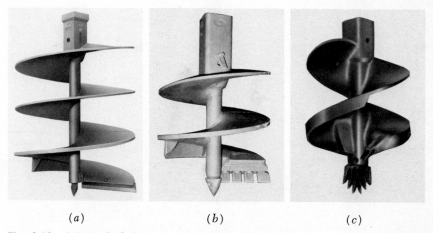

(a)   (b)   (c)

Fig. 4-11 (a) Single-flight auger bit with cutting blade; for soils. (b) Single-flight auger bit with hard-metal cutting teeth; for hard soils, hardpan, rock. [(a) and (b), *Watson Manufacturing Company, Fort Worth, Tex.*] (c) Cast steel heavy-duty auger bit for hardpan and rock. (*Hughes Tool Company, Inc., Houston, Tex.*)

(a)

(b)

Fig. 4-12 (a) A 45° underreamer ("belling bucket") for making a 21-ft underream in an 8-ft-diameter hole. This underreamer has hard-metal teeth for hard or stony soils, hardpan, soft rock. (*CALWELD, Division of Smith Industries International, Inc., Santa Fe Springs, Calif.*) (b) A 30° underreamer with blade cutters for soils that cut readily. Bucket diameter is 48 in., bell diameter 12 ft. (*Watson Manufacturing Company, Fort Worth, Tex.*)

(or other hard metal) teeth at the bottom edge. These teeth cut a small clearance both inside and outside the core barrel to facilitate flushing out cuttings, pulling the core barrel, and subsequently removing the core. The teeth are sometimes formed by building up the hard metal on a flame-cut, serrated edge of the core barrel, and sometimes are in the form of commercially available teeth which are set onto lugs formed on the bottom of the core barrel. In either case, the arrangement, spacing, and orientation of the bit teeth are important. The manufacturer usually furnishes detailed instructions as to the building up or setting of bit teeth.

"Calyx" or "Shot" Barrels For hard rock, which cannot be cut readily with the core barrel set with hard-metal teeth, a "calyx" or "shot" barrel can be used to cut a core of rock. In this barrel, the cylindrical skirt of the barrel extends upward some distance above the plate to which the drill stem is attached and which forms the top of the core barrel. The cutting edge of the barrel has no teeth or hard metal; it is composed only of the steel of which the barrel itself is made, with "feeder slots" cut into the skirt. The cutting is performed by chilled steel shot, which are poured into the hole, fed to the bottom through the slots, and are ground up under the rotating edge of the barrel. The grinding action of the broken steel shot grinds up the rock, and the fine steel dust and rock dust produced are washed into suspension by the water which is constantly pumped down the drill stem and which serves the dual purpose of keeping the grinding surface cool and carrying off cuttings. The suspended cuttings, rising through the narrow annular space around

Fig. 4-13 Rock core barrel ("coring bucket") with replaceable carbide-tipped teeth. (*CALWELD, Division of Smith Industries International, Inc., Santa Fe Springs, Calif.*)

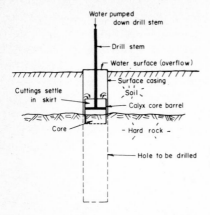

Fig. 4-14   Calyx or shot barrel, for coring hard rock.

the core barrel (clearance being very small), emerge from this annular space into the water-filled hole above the core barrel and settle into the skirt at the top of the core barrel. This skirt is the "calyx" that gives this type of core barrel its name. Progress with this coring is slow, but a "calyx" barrel can be used to cut cores of the hardest rock. Operation of the calyx barrel is shown diagrammatically in Fig. 4-14.

Multiroller-type Bits   For cutting large-diameter holes through hard rock, particularly where the holes have to be very deep, an adaptation has been made of the roller-type drilling bit which has been used for many years in the oil industry. The cutting elements in these bits are rollers furnished with teeth of hard metal; they drill by crushing the rock, producing small cuttings which can be suspended in drilling fluid and flushed out of the hole. A sufficient number of rollers is disposed over the face of the bit to cover the entire area of the bottom of the hole as the bit rotates. This type of bit is most often used with a "reverse circulation" type of rotary drilling rig, such as those used in the oil drilling and water well industries.

A drilling machine and roller bit with reverse circulation are illustrated in Fig. 1-5; a roller bit with "calyx" skirt, for use with compressed air, is shown in Fig. 4-15.

Air-operated Down-hole Bits   A drilling machine called the Magnum drill, manufactured by Ingersoll-Rand in the United States, has been available for the last few years. This machine has a drilling tool equipped with multiple air hammers, impinging on the bottom of the hole and receiving their air supply through the drill stem. The hammers are reciprocating in action, and the bit is rotated while they operate; the cuttings of rock are picked up by the moving column of air and blown out of the hole.

Sidewall Grooving Tools   To facilitate development of the shearing strength of the soil or rock for a pier which depends upon sidewall shear for support, tools have been developed for cutting grooves or corrugations in the sidewalls of holes in materials which will stand without caving.   One form of sidewall grooving tool is shown in Fig. 4-16.

Bailing and Cleanout Tools Pier-drilling contractors are usually equipped with "bailing buckets," which are used to remove water from the hole when the drilling is complete.   Some drilling buckets are so designed that they can be used also as bailers to remove most of the water in the hole quickly.   Drilling contractors are often equipped with air lifts which can be lowered into the hole and used to pump out

Fig. 4-15 Multiroller rock bit, with calyx-type skirt, for drilling with compressed air.   (*Hughes Tool Company, Inc., Houston, Tex.*)

Fig. 4-16 Shop-made sidewall grooving tool, attached to top of drill bucket.

Fig. 4-17 Mudding auger, for simultaneously advancing hole and mixing mud slurry. (*Watson Manufacturing Company, Fort Worth, Tex.*)

Fig. 4-18 Overburden bit (and mud mixing cylinder), for drilling in sandy overburden in mud-filled hole (reverse circulation). Bit diameter is 8 ft (244 cm). (*Girdler Foundation & Exploration Company, Clearwater, Fla.*)

all but a very small amount of the water. In case it has not been possible to seal off water from entry into the hole, down-hole pumps, which can be left in place, operating until the concrete pour is started, are often used.

Mudding Tools The mixing of prepared clay with water to make "drilling mud" is often a difficult and time-consuming task. If there is a great deal of it to be done, the contractor would be well advised to obtain mixing machinery such as that used in the oil fields. If it is only an occasional task, or if little mud is to be mixed at a time, the driller without a mud pit and a water jet can use the mudding auger shown in Fig. 4-17 for small-diameter holes. This auger has a tendency to become unstable when rotating at mixing speeds if its diameter is greater than about 3 ft.

For larger-diameter borings, the "mudding bit" shown in Fig. 4-18, which simultaneously cuts into the bottom of the hole and mixes drilling fluid and cuttings, has been found useful and is much more stable in rotation than the mudding auger.

4-1.6   Casing and Liners

A distinction must be made between *casing* and *liners*. Casing is welded steel pipe of substantial wall thickness ("line pipe"). It may be either temporary, left in place only to maintain a clean hole until concrete has been placed, then withdrawn; or permanent, becoming part of the finished pier. Casing must always be strong enough to resist the crushing forces that might be exerted on it by soil and water pressures before the concrete is placed. The importance of this consideration for large piers is pointed up by the fact that the resistance to lateral forces of a vertical cylindrical shell is inversely proportional to the *cube* of its diameter. There are several recorded instances of serious construction difficulties produced by buckling of large-diameter pier casings before the concrete was placed [4.29].

Commercially available sizes and wall thicknesses of "line pipe" are given in Appendix D. (Note that casing sizes are *always* specified by *outside* diameter.)

Because of its weight, steel casing has to be handled in moderate lengths. In some areas it is customary to use 10- to 20-ft lengths of casing, attaching successive lengths by lugs and pins or bolts. In other parts of the country it has become customary to use telescoping casing, starting with an oversized hole and using successively smaller diameters of auger and of casing as additional depths of hole have to be cased.

Liners are intended only to act as forms to retain the concrete of the shaft. They are not expected to resist crushing forces or to contribute to the strength of the pier. Sixteen-gauge, galvanized, double-riveted corrugated metal pipe ("CMP") is often used. Because of the corrugations, the *inside* diameter of CMP should be the nominal diameter of the pier shaft (or larger).

Other fabrications, such as lighter-gauge welded pipe and even Sonotubes, have been used under appropriate circumstances. Liners are protected from outside pressure by temporary casing or, where soil conditions are suitable, by being placed in a stable, noncaving, oversize shaft hole.

4-2   DRILLING AND CASING TECHNIQUES AND PRACTICES

4-2.1   Readily Drilled and Noncaving Soils (Auger Drilling)

Straight-shaft Drilling In soils in which drilled holes stand open readily, straight shafts are drilled very quickly, usually with an open-helix auger having two or three turns of a single flight. The auger may be equipped with either a knife-blade cutting edge or a series of cutting teeth (usually replaceable) faced with hard metal (see Fig. 4-11). In general, the auger with a blade is used in uniform soils that are soft to stiff or very firm in consistency, while the toothed auger is used in hard, cemented, stony, or very tough clayey soils that do not cut readily with the blade. Best progress is made with the blade when the auger is advanced fast enough to allow it to make a uniform cut but not to "corkscrew" itself into the soil like an earth anchor. A well-designed auger has a cutting blade on the side as well as the bottom to cut a clean hole and to make sure that the filled auger can be lifted without having to shear or tear the soil at its periphery.

Cutting blades and teeth should be kept sharp. Drilling with dull tools wastes time and money.

Use of Water in Auger Drilling Some clayey or slightly clayey soils, particularly where the water table is deep, drill readily but have a blocky or crumbly structure that leads to sloughing or caving. It is common practice, in areas where these soils are encountered, to add water as the hole is drilled. This produces a smooth layer of moist, cohesive, remolded soil on the sidewall of the hole, often allowing the pier to be completed without casing and without sloughing. This practice is acceptable only for end-bearing piers.

Machine Underreaming As mentioned elsewhere, any shape of under-ream—30°, 45°, domed, flat-bottom, stepped-bottom—is satisfactory if its shape is regular, its bottom is clean, and it stands without caving until concrete is placed.

The 30° underream is often specified in building codes—probably because, like specification provisions, building codes are often copied from each other without regard for the real significance of or need for some of their requirements. Some contractors feel that a 45° under-ream has a substantial advantage over the 30° one because its volume is substantially less; it can be cut much faster, thereby incurring less incidence of sloughing and caving if concrete is placed promptly; and it requires much less concrete. Another advantage claimed for the 45° bell is that, for the same bell diameter, the 45° tool is not as high as the 30° one; and consequently, a 45° tool can be used where a 30° tool would be too tall to clear the driving mechanism on the drilling rig.

Designers object that the edge stresses are higher in the 45° bell. Certainly this is true, but in most instances even in the 45° bell they are not high enough to endanger the footing, and they can be controlled by governing the height of the toe (the cylindrical section at the bottom of the bell).

The authors believe that, as long as the edges are not overstressed, one style of underream is as good as another so far as structural support is concerned.

In using the underreamer, the drill operator will establish a reference point on his kelly as soon as the tool rests on the bottom of the hole, to enable him to identify the true bottom on each trip into the hole. The underreamer is designed to complete its action in several "trips," but without deepening the hole or disturbing its bottom. In an operation where there is a tendency to pack cuttings under the underreamer, creating a false bottom of remolded soil, it may be necessary to clean out and reestablish the true bottom several times during the underreaming operation.

Some underreaming tools are made with a "reversed cutting" blade in the bottom of the bucket, which slides on the bottom of the hole during normal underreaming operations and picks up soil from the bottom when the reaming blades are retracted and rotation is reversed, thus permitting the true bottom level to be reestablished when necessary.

Underreaming cutters are activated by pressure from the kelly through a lever system housed in the underreaming bucket. If the drill operator tries to take too large a load into the bucket, progress will be impeded in several ways: retraction of the cutting blades is made difficult, spoils will be packed into the bucket so tightly that their removal is impeded and time is lost, and complete plugging of the bucket may make it necessary to pull out against a partial vacuum in the hole. An attempt to lift out an overloaded bucket that has the reaming blades not completely withdrawn will result in raveling the sides of the shaft, or hang-up on the bottom of the casing if casing is present.

Best production is usually obtained by coming out and emptying the bucket as soon as it is about one-third filled with loose—not packed—spoil.

Some soils will stand open nicely in a straight hole, but cave and slough when an attempt is made to underream them. In this case, it is sometimes economical to drill a straight shaft to the full underream diameter (or nearly full diameter), then cut a slight underream, pour a footing at the bottom, and set dowels and a corrugated metal liner of specified shaft diameter. The annular space outside the liner can be filled with sand, soil, or lean concrete, and the liner then filled with concrete of the required strength. When this technique is used, consideration must be given to the possibility of subsidence of loose backfill, especially if the pier is adjacent to a slab-on-ground floor.

Manual Construction of Underreams In some soils and weathered rocks that tend to fracture, ravel, or cave when the rotary underreaming tool is used, and in hard rocks, when an underream is required, hand excavation or pneumatic tools may be used [4.3]. Hand excavation also may be used in readily drillable soils when the footing is to be of larger size than the largest available underreamers [4.5, 4.6]. In some cases, the excavation will stand open safely for this operation, but more often (for example, in moist sand) shoring and bracing must be installed (or "spiles" driven) to maintain the roof as handwork proceeds. In the Los Angeles area, it is reported that hand-cut underreams are common for diameters of more than 10 ft (3 m), and that the 1-in. by 4-in. wooden shoring is commonly left in place when the pier is concreted.

Multiple Underreams Although rarely used in the United States, multiple underreams are sometimes used for the purpose of spreading the pier load over a wider zone in stiff soils or in layered soils where there are two or more strata of superior bearing capacity. Figure 4-19 shows a typical arrangement of two underreams, designed to mobilize the shearing resistance of soil over a larger area than would a single underream at the bottom. Additional advantages claimed [4.7] by its proponents for the multiple-underream pier (over a cylindrical pier offering the same sidewall shear area) are: (1) that because the shear is developed in undisturbed soil rather than in a softened "skin" adjacent to the concrete, and no reduction factor for softening or "smear" need be applied, the load-carrying capacity of the pier is increased about

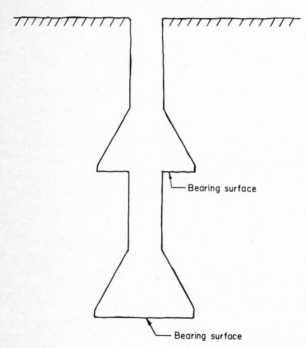

Fig. 4-19   Multiple-underream-type drilled pier.

50 percent; and (2) that the volume of concrete for the multiple-underream pier is only 30 to 40 percent of that required for the same diameter straight shaft.

Any contractor who can make piers with an underream at the bottom can drill the multiple-underream configuration—provided the soil is suitable for underreaming and stands without caving at the selected levels. Choice of underream levels and diameters is a design problem, the responsibility of the geotechnical engineer for the project, and may be open to serious question. The actual feasibility of this type of construction will have to be proved in the field for each project.

Drilling Battered Holes   Battered pier holes are frequently required to provide horizontal reaction components for either compressive or tensile loads. Battered holes for nonfoundation purposes—for drainage or for tieback anchors, for example—are not uncommon. Most truck-mounted drilling rigs designed for auger drilling can operate with the kelly at any angle between vertical and horizontal (and some even above the horizontal). Many crane-mounted machines can drill at batters up to 45° (see Appendix A). Truck-mounted rigs designed for drilling with the drill bucket rather than the auger do not usually have the ability to drill with the mast tilted back. With these machines, batter is obtained by tilting the entire machine, either longitudinally or to the side, and for practical purposes the obtainable batter is limited to about 1 horizontal to 3 vertical.

The problems of drilling a battered pier hole are not confined to the setting of the angle of the rig mast and kelly. Setting the angle with the vertical is usually a simple matter, and can be done quite accurately. Setting the direction angle of the batter (i.e., the azimuth, or directional bearing) is not so simple. The vertical angle will be indicated by a plumb bob and scale mounted on the rig mast. The direction, if it is to be set with any precision, must be determined by observation of the direction of the tilted mast by means of surveying instruments. Fortunately, a variation in the directional bearing of a battered hole is (in the authors' opinion) not usually of great importance, and a rough compass bearing—or even an "eyeball" setting—will generally suffice.

There may be a tendency for a battered hole to change its slope as it is drilled. If the soil formation is homogeneous (a rare condition!), there will be a tendency for the angle of the hole to become less steep as penetration proceeds, because of sagging of the unsupported middle of the kelly or drill stem. In layered formations, the deflection might be either upward or downward (or to either side), depending on the orientation and sequence of harder and softer strata. Fortunately, such deviations from straightness are usually of no structural significance for drilled piers or anchors. On the other hand, deviation of horizontal (or sloping) holes is extremely important when the hole is intended to tap an underground water source. This is true regardless of whether the objective is to intercept a natural aquifer or to tap a series of inter-connected vertical wells designed to intercept groundwater flow. Alignment of such holes is a continuing problem for both geotechnical engineer and contractor.

It is important that design drawings and specifications should be reviewed for unnecessarily restrictive requirements for battered holes before a contract is signed. The drilled pier contractor should reconcile what he *can* do with what he *is required* to do before he commits himself. A designer who believes that his tolerances on batter, direction, and straightness are the maximum that can be allowed may be quite willing to permit larger tolerances in exchange for a larger-diameter hole or underream, or a deeper hole, or an increase in reinforcing.

In drilling a battered hole, there is more tendency toward caving than in vertical holes in the same formation; and consequently, casing requirements may be more critical, and underreams may not be as feasible as for vertical piers.

The concreting of a battered hole with an underream (or bell) at the end may introduce special problems. Because of an increased tendency to cave, the hole may have to be concreted more promptly than a vertical hole. The positioning of the reinforcing in the shaft is more difficult when the batter is very flat. Higher-slump concrete may be required, and the geometry of the underream may be such as to prevent complete filling with concrete, unless a special pipe is installed to vent trapped air in the upper part of the underream (see Fig. 4-20).

4-2.2   Hard Drilling

Boulders, Cobbles, "Hardpan" For hard or stony soil, auger bits and underreamers are usually set with hard-metal-faced ripping teeth rather than with cutting blades (Figs. 4-11, 4-12). These are very rugged and, used with a drilling machine having sufficient torque, will drill most cobbly soils, hardpans, and soft or decayed rock with impressive speed. Hard boulders, however, will stop any auger. If a stone is not more than one-third the diameter of the hole, it may be possible to pick it up with the drilling auger, or with a special short auger bit called a boulder extractor (Fig. 4-21), or sometimes with a "grab" (Fig. 4-22). If a stone is too large or too tightly held to be picked up thus, it must be broken up or loosened first. This can often be accomplished by dropping the kelly on it, or the kelly equipped with a "gad" or "spudder," or it can be loosened (and sometimes broken)

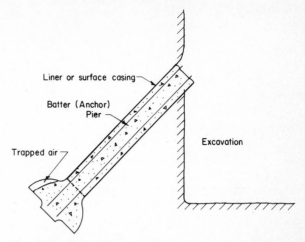

Fig. 4-20 Battered pier, showing how air can get trapped during concreting.

Fig. 4-21 Boulder extractor, a tool for picking up cobbles, small boulders, or broken ledge rock. (*Watson Manufacturing Company, Fort Worth, Tex.*)

Fig. 4-22 A hydraulically operated "grab," "rock bucket," or "cleanout bucket." This tool handles solids or fluids, and is used also for cleaning mud, laitance, and diluted or contaminated concrete from the top of a completed column, as in the illustration. (*W-J Sales Engineering Co., Inc., Houston, Tex.*)

by the use of a "boulder rooter"—a wedging tool twisted by the kelly (Fig. 4-23). Loosened or broken pieces can then be picked up by an auger or a grab, or even be extracted by hand or broken further by a man with a jackhammer.

Layered Rock, Cemented Soil When the auger is stopped by relatively thin layers of hard rock (often interlayered with soil or very soft rock) or by soil layers that have become cemented, the techniques described above for boulders can be used to break up the layers so that they can be drilled with the flight auger (equipped with hard-metal teeth).

Drilling in Rock—Rock Sockets—Coring When pier holes have to be drilled into shale or weathered rock, penetration can often be effected with a flight auger having hard-metal cutting teeth. This requires skill and experience on the part of the driller. Too much pressure ("crowd") can break the auger teeth, and excessively fast rotation can overheat and burn the teeth. With the correct speed and pressure the teeth will cut rather than abrade the rock.

When penetration cannot be effected with the flight auger, a core barrel or other special tool is required.

Except for use in very hard rocks, the core barrel usually consists of a simple cylindrical barrel set with cutting teeth which are faced with tungsten carbide or some other commercial hard metal. The arrangement and the spacing of the teeth are important; they must be so set as to cut sufficient clearance inside and outside the cylinder wall to accommodate the rock cuttings, they must be arranged so as to give complete coverage of the surface to be removed, and they must be in the right number to permit application of the necessary load per tooth through the drilling machine used. A core barrel of this sort cuts out an annulus, leaving a pedestal of rock within the barrel. Very often the core will break off before the capacity of the barrel is reached.

Fig. 4-23 Boulder rooter, for loosening or breaking up embedded boulders. (*Watson Manufacturing Company, Fort Worth, Tex.*)

If it does not, the barrel is removed, a wedge is driven on one side of the core to break it off, and the core barrel is lowered again over the loosened core. The core can usually be counted on to jam in the core barrel so that it can be lifted out of the hole, and can often be removed from the core barrel by tapping or hammering on the outside of the barrel while it is suspended.

When drilling with a core barrel of this kind, it is necessary to keep several feet of water in the hole at all times. Speed of rotation should be governed carefully to comply with the core barrel manufacturer's recommendations (usually somewhere between 10 and 20 rpm). The pressure applied depends on the type of rock being cut, and the best pressure is usually found by experimentation. Too much pressure will usually result in excessive wear or tooth breakage. With the right machine, a good well-set bit, and a good operator, it should be possible to cut 30- or 36-in. (75 to 90 cm) core at a rate of 2 or 3 ft per hr even in hard rock.

The coring of the rock by hard-metal teeth should be a cutting rather than an abrading process. To accomplish this the driller should allow the teeth to cool by lifting the bit every few revolutions. The core barrel is, of course, kept rotating during this brief cooling, and the raising and lowering of the core barrel serves the additional function of pumping cuttings out of the annular groove. Core barrel teeth must be kept sharp. The manufacturer will furnish recommendations for sharpening and building up worn teeth. Continuing to attempt to drill with a core barrel with worn teeth is inefficient, and has the additional disadvantage that whatever footage is made during this period is drilled without sufficient clearance and will give trouble when a core barrel with new or reset teeth is inserted.

Where the rock is too hard to be cut by tungsten carbide teeth, as is sometimes the case, a "calyx" or "shot" barrel, with chilled (hardened) steel shot as the cutting medium, can be used successfully (Fig. 4-12). The "calyx" barrel has to be removed and emptied of its cuttings at appropriate intervals. This technique of drilling with chilled shot is slow, but can be used to core the hardest rocks. It will not work where rock is broken or fissured and the shot can be lost into cavities.

A novel and very effective technique of accelerating the penetration of a "shot" barrel has been observed by the authors. When very hard rock is reached, a vertical hole, say 2 in. (5 cm) in diameter, is put down by jackhammer or by means of a small diamond core drill in the exact line of the annulus to be drilled by the shot barrel. This hole is then filled with chilled shot, and the drilling is started and proceeds in the usual manner. The presence of a fresh supply of the steel shot always at the cutting edge greatly expedites cutting action. (It

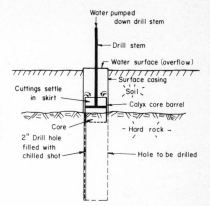

Fig. 4-24   Calyx or shot barrel, coring hard rock. A small hole, air-drilled down the edge of the core to be cut, and filled with chilled shot, is very effective in expediting the coring of very hard rock.

might be mentioned that this technique works very successfully in drilling through concrete with heavy steel reinforcing bars.)   This technique is illustrated in Fig. 4-24.

In addition to the techniques described, there are a variety of special tools and machines available for coping with difficult rock conditions. One of these is the Magnum drill, in which the cuttings are blown out of the hole by compressed air. This tool has a tendency to "gum up" when used in rock which breaks down into clay or is interlayered with clay. A disadvantage of this particular tool is that it requires enormous amounts of air, makes a great deal of noise, and showers the surrounding area with dust and cuttings, or, when drilling below the water table, with a mixture of mud and cuttings.

Jackhammer ("Dental") Work   When a pier hole is centered over an irregular rock surface, so that one side of the bit bears on rock and the other on soil, or is over a cavity, it becomes impossible to drill a straight hole. Despite this difficulty, drilled piers with rock sockets may be the most feasible type of foundation, and in this case machine drilling has to be supplemented by putting a man down the hole with a jackhammer. In some cases it will be possible to take out just enough rock to establish a level surface that can be drilled by core barrel. More often, it proves necessary to complete the rock socket or underream by jackhammer work, supplemented, in some cases, by blasting. The use of explosives under these circumstances requires expert planning and performance to avoid risk of damage to other piers, pier holes, or bearing surfaces.

This type of rock excavation in a pier hole is very expensive, with costs sometimes amounting to $300 or $400 per cu yd of material removed (1971). Rock conditions requiring such "dental work" are common where bedrock is of solution-prone limestone or dolomite.

4-2.3   Drilling with Roller Bits (and Other Nonauger Types)

Although a very large proportion of the pier holes drilled in the United States are drilled with auger-type machines, supplemented in hard rock by the use of core barrels and sometimes manually operated pneumatic tools ("jackhammers"), in some regions ground conditions are such as to make an entirely different drilling technique more favorable. A large part of the state of Florida falls in this category. In this area, a typical condition is an overburden of sand, a high water table, and a differentially cemented, fossiliferous limestone locally called "limerock." The sand is cohesionless, and often loose in condition. The limerock is often soft, porous, cavernous, and weakly cemented. For foundations that must withstand heavy lateral loads or uplifts—for example, electrical transmission towers and the like—it is desirable to set the foundations well into rock to develop substantial sidewall shear.

These foundations are usually of the straight-shaft, "rock socket" type. In this region it has been found advantageous and has become customary to drill the holes, in diameters up to 60 in. ($1\frac{1}{2}$ m) and to depths that could, if necessary, be more than 100 ft (30 m), using rotary drilling equipment and techniques adapted from oil field practice. The Wabco 1500 heavy-duty drill, shown in Fig. 4-8, is typical of the drilling machines used. In cohesive soils, the machine may be used as an auger machine (although it is not as efficient as machines designed for that service); in cohesionless soils and below the water table, drilling is done in a mud-filled hole. A special "mudding bit" (Figs. 4-17, 4-18) is used to make the hole and to keep the drilling mud mixed and the cuttings in suspension. (See p. 141 and Appendix B for discussion of the use of drilling mud.) When rock is encountered, the bit used is the multiroller type (Fig. 1-5). The toothed rollers break up the rock into fine cuttings, which are carried out by the drilling mud as it circulates. Except for piers through water (p. 144), piers installed by this technique do not usually require casing. Because their concrete is placed by tremie, displacing the drilling mud, they are designed for support by sidewall friction only, with no allowance for end-bearing. While the sidewalls are not subject to direct inspection, experience and full-scale load tests have shown that good bond for shear transmission can be developed between the concrete and the rock. An anchor pier for a television tower constructed in this manner is shown in Fig. 4-25.

4-2.4   Sidewall Grooving

The degree of transmission of downward forces to the sidewalls of a pier shaft or rock socket has often been questioned. Many tools have

(a)

(b)

Fig. 4-25 Tower and anchor piers for 1,500-ft TV tower near Tampa, Florida. (a) Reinforcing cage being placed.   Pier contains 2 percent steel.   (b) Finished anchor.

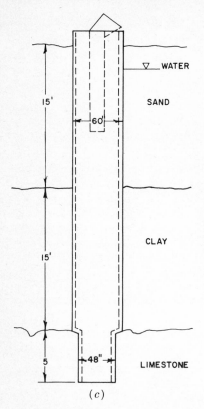

Fig. 4-25 (Continued) Tower and anchor . . . Florida. (c) Dimensions of anchor pier. Tower pier was identical except for top detail. (*Girdler Foundation & Exploration Company, Clearwater, Fla.*)

been developed for grooving or roughening the sidewalls of shafts or sockets drilled both into soil and into rock. Because of the designers' uncertainty regarding shear transmission between pier and soil or rock, the use of these tools is often specified in some areas of United States practice whenever the design used depends upon pier support by side-wall shear. Application of this technique seems to be a matter of local geology, local experience, and local availability. In the Denver area, for example, such grooving of pier holes drilled into rock is considered routine. In northern California, straight-wall piers drilled into hard soil are sometimes specified [4.33]. In some areas, this technique is considered technically unjustified and a contractor's nightmare, and is not used at all.

4-2.5 Squeezing Ground

Occasionally, when a hole is being drilled in otherwise competent ground, a soft stratum, usually of clay or organic silt, will be encountered which tends to squeeze in and reduce the diameter of the hole while the drilling is being performed. To permit this action to continue while the hole is being deepened is to risk loss of ground and delays in drilling due to hang-up of the auger bit. Whenever a squeezing condition is encountered, casing should be inserted and used to limit this action as far as possible. The casing may have to be driven or drilled ahead of the excavation to get it past the zone of squeezing soil. The twisting bar shown in Fig. 4-26 can be used for this purpose.

It should be remembered, however, that even though firm soil or rock is encountered below, and drilling may proceed below the cased-off zone, the squeezing soil will seal to the casing and may give trouble later due to sealing off of lower groundwater, or to the squeezing forces being in excess of concrete pressure at the time the casing is withdrawn. A good rule would be to leave the casing in place in such formations.

4-2.6 Caving or Sloughing Soils

Cohesionless soils (gravels, sands, silts) will usually cave when drilled by an auger either above or below the water table. Exceptions to this occur when such soils are cemented, or when they contain enough moisture to establish capillary forces but not enough to completely saturate them. Gravel will not be restrained by capillary forces unless it is intermixed with or has its surfaces coated by fines of suitable particle size to develop the necessary capillary forces. Silty sands and well-graded sands, if moist but not saturated, will often stand without caving, at least for limited periods. So will fine sands and silts, but these are more likely to be saturated; and if nearly saturated, they can be liquified by vibration from the drilling, with the result that the hole collapses completely.

Fig. 4-26 Casing twisting bar. This tool, attached to the kelly, is used for "screwing" down casing through squeezing or caving soil, for seating casing into rock or clay stratum, or for pulling casing. (*Watson Manufacturing Company, Fort Worth, Tex.*)

Casing the Hole   The most usual and direct means of coping with caving cohesionless soils is to case off as the hole proceeds through the caving formation.   This can be a tricky operation.   It is usual to drill the hole 2 in. larger in diameter than the OD of the casing to be used.   Even in caving soils, it is often possible to drill the hole quickly and insert the casing freely before caving has started or reached troublesome proportions.   But conditions will be encountered where this cannot be done.   When pore-water pressure is sufficiently high, sands and silts cannot be removed to extend the hole past the bottom of the casing without causing the soil to "run in," and serious loss of ground may result.   Under these circumstances it may prove impractical to struggle with alternate drilling and driving of casing (for driving or twisting will be necessary whenever the caving ground seizes the casing).   Several techniques are available, however, one or another of which may serve to keep the hole open while drilling is being performed and allow the casing to be either advanced with the hole or dropped in freely when the shaft hole is completed.   The first of these is to keep a positive head of water in the hole during drilling—i.e., keep the hole full of water, or nearly enough full that hydrostatic pressure is sufficient to prevent run-in of sand at the bottom—and to alternate drilling or screwing the casing down ahead of the boring, with drilling (usually with the bucket auger) to remove the soil within the casing (but not ahead of the casing).   A second method is the use of a drilling mud slurry in the hole to balance the hydrostatic pressure of ground-water outside.   A third is to draw down the groundwater level around the hole, usually by the use of deep wells suitably arranged and pumped down to a level well below the bottom of the proposed drilling.

When casing is practicable but installation is difficult because of caving or seizing formations, the use of a vibrating pile driver is sometimes very effective, particularly in granular soils.   In clayey soils, the casing will sometimes build up a cake of clay both outside and inside the casing, so that penetration is greatly slowed after a few insertions and removals.

Stabilization by Drawdown   The drawdown procedure often works well in fine and well-graded sands.   As water level is drawn down *outside* the hole, seepage forces around the periphery of the hole are reversed; capillary forces are set up which tend to keep the sidewalls from caving; and, if drilling is performed smoothly and without delay, a clean open hole can often be established all the way through the caving stratum.   Suitability of soil conditions for this technique can be established from the test borings, correlated with observations of the actual pier drilling as it proceeds.   Where soil conditions are favorable, it may be possible to greatly expedite a pier-drilling job by the use of groundwater drawdown, as compared with the best progress that could be obtained if casing were used to seal off the caving strata. However, use of this technique can produce substantial ground subsidence, and this possibility should not be overlooked.

Stabilization by Mud Slurry   The use of a drilling mud slurry to keep the hole open can be applied to a wide variety of soil conditions. This technique has been adapted from oil well practice.   A discussion of drilling mud—what it is and how it works—is given in Appendix B.   For example of its use see Refs. 4.5, 4.6, 4.8, 4.11.

In some parts of the country, notably those in which oil well drilling operations are currently going on, it is possible to have drilling mud, in slurry form, delivered to the job site in tank trucks.   Elsewhere, the contractor who uses the technique will have to mix his own slurry and will need some special tools to do it, as well as special tools for drilling in a mud-filled hole (Fig. 4-17, 4-18).

As drilling proceeds in a mud-filled hole, additional slurry must be furnished to replace the volume of the spoil removed, and inevitably some slurry is removed with the spoil.   Slurry to keep the hole filled is usually stored in a "mud pit" excavated adjacent to the hole, feeding into it by gravity through a ditch; or it may be kept in barrels or tanks and discharged into the hole as needed.

To a person not accustomed to its use, the efficacy of a viscous or heavy drilling fluid to keep a hole from caving is surprising.   It will allow drilling a hole in sand—dry, moist, or saturated—with straight, clean walls.   It can even enable the driller to underream in a clean sand below the water table, without any caving or sloughing of the roof of the underream.   This performance is not automatic.   It requires skill and experience on the part of the driller—and understanding of the technique and confidence in the contractor on the part of an engineer who authorizes or approves its use.   Whatever drilling is done in a mud-filled hole is done blindly; the results can be judged only by behavior of the tools, augmented by probing.   The sidewalls and underream cannot be seen, even with the use of an underwater television camera.   It would be possible, of course, to have the surfaces inspected by touch, by a diver, but the authors know of no instance of this being done.   A final, indirect check on the hole can be made by comparing its design volume with the volume of concrete that is placed in it.

When starting a hole that is to be mudded (as well as holes under many other circumstances), it is advantageous to begin by drilling an oversized hole and setting a short (8- to 15-ft)(2½- to 4½-m) piece of oversize surface casing.   This helps to keep a clean hole, preventing caving and fall-in at the surface caused by operations of men and machines.

In instances where the bottom of the hole (or the underream, if any) is to be in a noncaving material, the hole should be drilled oversize down to the lower stratum. Casing somewhat larger than the pier shaft can then be set through the slurry and seated into the lower stratum, so that the slurry can be bailed or pumped out. The remainder of the hole can then be drilled in the dry (which is much faster than drilling in slurry). A liner is then seated in the top of the uncased part of the hole to retain the concrete of the pier shaft when the casing is pulled. After the concrete has been placed, the casing is pulled, taking care not to disturb the pier shaft. This leaves an annular space around the liner, filled with slurry, which should then be filled with something more solid to give the shaft lateral support and inhibit buckling action under load. Filling can be done by pouring in pea gravel, which settles to the bottom, displacing the slurry and causing it to overflow at the top.

Experiments have shown that concrete properly placed in a slurry-filled hole makes a good bond with reinforcing steel, despite the thin coating of slurry that wets the reinforcing bars. Designers often question the reliability of the bond established with the sidewalls of a hole in soil or in rock (for development of sidewall shear). Recent experiments indicate that a very good bond can be developed. Where the drilling of pier holes designed for sidewall shear is common practice, it is not unusual to permit the use of drilling mud provided the required pier capacity can be demonstrated by full-scale load tests on a designated number of completed piers.

A recent development in drilling fluids is Johnson's "Revert." This is a material which, added to water, makes a readily pumpable but viscous fluid that will carry out cuttings as effectively as a mud slurry. After a predetermined time, however, it reverts to about the viscosity of water, and is therefore very easily displaced by concrete. This material has had extensive use in the water-well-drilling industry, but has had little reported use (as of 1971) in the civil engineering field [4.32].

4-2.7 Sealing Casing into Rock

When a pier is to be founded on rock or drilled into rock, there is often difficulty in making a satisfactory water cutoff at the rock contact. Where the rock is soft or weathered, a good seal can usually be effected by driving the casing, or by twisting or rotating it while pushing it down. If the plain end of the casing will not penetrate far enough to make the seal, flame-cut teeth on the bottom may be effective for cutting into soft or weathered rock, and the cuttings formed will help in establishing the water seal. For harder rock, it may be necessary to face the cutting teeth with hard metal [4.9, 4.10].

Under some conditions sealing of the bottom of the casing may be extremely difficult or not even possible with the equipment available. This occurs when the rock is hard and the surface is sloping or irregular, or when there is a layer of cobbles or boulders immediately over the rock surface, or when the bedrock is fissured or cavernous. Under such circumstances, grouting with cement or chemicals might be considered, with the objective of forming an impervious bottom pad which could be drilled out later. One large pier-drilling contractor reports successful sealing under such circumstances by using a proprietary material, MECO "Sealite" (Maintenance Engineering Company, Philadelphia), mixed with cement and packed into the leaking crack, leaving a small pipe in the crack for pressure relief. The relief pipe can be plugged or grouted later.

Alternatively, attempts to seal might be abandoned, and tremied or pumped-in concrete be placed in the water-filled hole (see Art. 4-5).

4-2.8 Hole Location and Plumbness

Most pier-drilling rigs are so built that it is easy to center the auger over a stake marking the pier location, and the "stinger"—a small pilot point or bit on the bottom of the auger—keeps the auger centered as the hole is started. However, when the hole must be started in hard, broken, or sloping rock (or even sloping hard ground), it may be difficult to keep the bit from "walking off" as the driller attempts to start the hole. In such cases, a pilot hole may be drilled first. In other instances, especially where the hole has to be started with a core barrel or at a batter, a short, cylindrical starting guide can be set to keep the bit or barrel from "walking off." Starting guides are available from some drilling-machine manufacturers, or they can be readily improvised in the field.

Vertical alignment (plumbness) is usually easily maintained, for the auger and kelly are very heavy and there is little tendency for a hole to "wander off" as it is drilled. Some machines depend on the weight of bit and kelly alone for "crowd" (downward force), governing the application of this force by regulating the tension in the wire line from which the kelly is suspended. Other machines have a hydraulic "crowd" mechanism to augment the weight of lighter tools and the kelly. Probably somewhat more care should be taken in the latter case to check plumbness of the hole, particularly in hard ground. The authors do not believe that this consideration is often of importance, because the deviation from plumbness that occurs under usual conditions is small and is of no significance as far as pier performance is concerned.

The greatest threat to plumbness is the presence of boulders or rock pinnacles. Where such obstructions produce serious deflections of drilling tools, corrective work, usually by jackhammer, is required. Under these conditions, the completed hole is often considerably oversize.

4-2.9   Piers through Water into Soil or Rock

Drilled pier foundations are frequently used for waterfront structures, bridge piers, and other construction in shallow water. When soil conditions below water are suitable for the use of this type of foundation, casing can be drilled or driven down to where it is firmly held by the soil, and then drilling can proceed just as for an installation on land below a high water table [4.11, 4.12]. Where the casing can be sealed into a suitable impervious formation, it can be dewatered and the drilling and underreaming can proceed in the dry; or, if dewatering proves impractical, the mud-drilling technique (p. 141) can be used to complete the pier hole. A discussion of concrete placement in such cases is given on p. 148.

4-2.10   Underpinning

Drilled pier–type foundations are particularly adaptable to underpinning problems for several reasons. Pier holes can be drilled immediately adjacent to existing foundations, with relatively little vibration and without ground displacement. Pier construction is rapid, so that underpinning construction can be completed in a minimum of time. Under most circumstances the underpinning pier can be taken down to a safe bearing level with comparative ease. In some instances, underpinning piers can be combined with drilled pier diaphragm walls, forming part of the basement wall for the adjacent new structure. For interior underpinning work, as in remodeling or in restoration of support to settling interior columns, specially mounted pier-drilling machines for operation under low-headroom conditions are available in many cities.

Precautions must be taken against loss of ground during drilling operations. When squeezing or caving ground is encountered, careful use of casing may be required; under some circumstances, the use of drilling mud may be adequate to prevent caving or hole collapse, even when the soils drilled are cohesionless and under the loading conditions produced by the foundation being underpinned. When the underpinning shaft is drilled through stiff clay, the hole will usually stand open long enough to permit completion and concreting before appreciable ground movement can develop.

4-3   PREPARATION OF BOTTOM OF HOLE

4-3.1   Cleanout Techniques

Any pier hole which depends upon bottom-bearing for its support must be cleaned out carefully and completely before concrete is poured. It is possible with some auger bits, and in some soil or rock, to drill a clean bottom. However, in most underreamed holes and in many straight shafts, cuttings or "slop" on the bottom of the hole may require handwork to produce a good bearing surface. This necessitates descent of a workman into the hole, with the appropriate cleanout tools, and a subsequent descent by the inspector to see that the job has been done properly.

4-3.2   Shape of Bottom of Hole

Different tools cut bottoms of different shapes. Some bottoms are nearly flat, some are dished, some have a small pilot hole in advance of the main hole, and some are rounded or conical in shape. As long as the hole is cleaned out properly, the shape of the bottom makes no appreciable difference to the bearing capacity of the pier. In the authors' opinion, there is no objection at all to the use of tools which cut a "stepped" bottom, or which make a pilot hole a few inches deeper than the bottom of the pier.

4-3.3   How Clean Is Clean Enough?

Cleanness is a question which is usually left up to the judgment of the inspector, but it is the authors' opinion that the inspector is rarely qualified to exercise judgment in this matter. If there is difficulty in getting a clean bottom in a pier hole, the matter should be called to the attention of the geotechnical engineer at once; he should make an inspection, and he should give definite instructions as to what should be done. A small volume of dry or plastic cuttings on the bottom of the pier hole will make no difference in bearing capacity. If, however, there is as much as an inch or two of soft "mud" in a plastic condition, which will not be displaced to the outer circumference as concrete is poured, but which is compressible enough to permit measurable settlement, then this would be unacceptable, and special means would have to be devised for getting the hole cleaner. It must be remembered that, where the bottom of the hole is in very impervious soil such as clay, the cuttings or mud trapped under the concrete as it is placed may not have an opportunity to compress appreciably in the brief time that it takes for the concrete to set up, and therefore they could constitute a compressible layer which might produce measurable postconstruction settlement.

Cleanup of sidewall smear is, in the authors' experience, rarely needed or required.

It is sometimes the practice to "stabilize" the bottom of a hole containing a few inches of water, mud, or "slop" by adding enough cement, or gravel-cement mix, to absorb the water and form a paste or soil-cement mix so rich in cement that it will set up sufficiently to be practically incompressible under load. To form such a mix, the added dry materials will have to be thoroughly blended with the material in the hole. This may be attempted by inserting the auger in the hole and mixing by a few turns of the auger, without allowing the blade to drill any deeper. The authors have seen this method used, and do not know of any postconstruction settlement that can be attributed to it; but they regard the practice as questionable. If this in-place bottom stabilization is used, it must be done carefully; and it should be avoided whenever possible. The use of a smaller sump filled with "stabilized" material is preferable.

In instances where bottom softening is due to seepage of small amounts of water from above, a "mud slab" (a few inches of concrete placed immediately on completion of the boring) will protect the bottom soil and facilitate later cleanout just before the main mass of concrete is placed. But when water is rising under pressure from below, a mud slab would probably be lifted or broken by water pressure and could be responsible for later sudden settlement.

For piers designed for sidewall shear support only, bottom cleanout is of minor importance. In such cases, the only cleanout is performed with the drilling tools, and manual cleanout and visual inspection of the bottom are often omitted entirely. Of course, care must be taken that enough sidewall contact for the concrete is available above any disturbed material remaining on the bottom, and that whatever "slop" is present is not fluid enough to intrude or dilute the concrete above the level at which shear support is counted on.

4-3.4 Bad Air or Fog in the Hole

Bad air or poisonous fumes can come from many sources, and for safety reasons this possibility must be taken into account whenever any person enters the pier hole (see Art. 5-1: "Safety Considerations—Entering the Hole"). A common occurrence, particularly during winter construction, is the formation of fog in the hole, so that inspection of the bottom is impeded or prevented [4.13, 4.17, 4.29]. When either bad air or fog is present, it should be dispersed by pumping in air before and during descent into the hole by any person.

4-4 DEWATERING

To assure good bearing and high-quality concrete in footing and shaft, a pier hole should be essentially free from water when concrete is placed. The technique of sealing off water with casing has been described in Art. 4-2. In instances where water has entered the hole before the casing seal was effected, the hole can be bailed almost dry by the use of a bailing bucket attached to the kelly. Final mop-up is usually completed by a man down the hole using a bucket and a scoop.

In other instances, a small amount of water in the hole can be taken care of by drilling a sump hole in the bottom of the pier hole, pumping down until only a few inches of water remain in the sump, then filling the sump with clean gravel or with a mixture of gravel plus dry cement, proportioned to allow the bottom water to not quite fill the voids. The concreting can then follow in the normal manner. The presence of a small plug of very lean concrete, or even of saturated clean gravel, below the center of the pier will not reduce bearing capacity or increase postconstruction settlement.

Where water cannot be shut off completely, it is sometimes feasible to use a pump fitted with a length of small-diameter pipe on the intake, keeping the pipe on the bottom and the pump running until concrete placement has started. The pump can be lifted out after the concrete head is high enough to prevent further entry of water into the hole.

A small amount of water left on the bottom of a pier hole can produce a surprising volume of clean gravel instead of concrete at the bottom of the pier. Larger amounts can be quite disastrous when the concrete is placed by free fall. There is one report of 2 ft (60 cm) of water in the hole resulting in 8 ft (2½ m) of sand and gravel at the bottom of the pier [4.17].

4-5 CONCRETING

4-5.1 Placement of Concrete

Prompt placement of concrete in the prepared hole, as soon as it is completed, is a necessity if troubles from caving, squeezing, water entry, and difficulty in pulling casing are to be avoided.

The manner in which the concrete is placed is also of the greatest importance.

Free Fall Experience and experiment have shown that concrete falling freely does not segregate provided the falling concrete does not strike the sides of the hole or a reinforcing cage [4.14, 4.16]. McKinney Drilling Company advises that in the pier foundations for the new Federal Building, Cleveland, Ohio, free fall of as much as 167 ft (51 m) was allowed.

It is not sufficient (except in very shallow holes) to discharge the concrete directly into the hole from the chute on a mixer; this will almost invariably result in the concrete's striking the side of the hole (or the cage) as it falls. And the authors do not consider satisfactory the common practice of having a workman hold a shovel so as to deflect the concrete stream from the chute so that (maybe) it falls straight down instead of striking the cage or the side of the hole.

It is necessary to have a hopper centered over the hole, with a bottom spout designed to concentrate the falling concrete in a stream of small diameter compared to the hole diameter, or else to place the concrete through a tremie pipe or "elephant's trunk" discharging at or near the bottom of the hole. In order to completely eliminate segregation, the hopper should never be emptied until the hole is filled with concrete. This is impracticable when dealing with large piers requiring several loads of ready-mix, or when casing must be withdrawn in sections; but the interval between discharges should be kept as brief as possible to avoid the development of a layer of laitance on top of the concrete in the hole before the next truckload of concrete is placed (refer to p. 155, "Interrupted Pour").

Tremie Placement; "Elephant's Trunk"; Pumped Concrete In many piers where reinforcing cages are used, the cage diameters will be small enough to obstruct the free fall of concrete. Where inside diameter of cages is less than 2 ft (60 cm), tremie placement can sometimes be used, but pumped-in concrete may be required.

Tremied or pumped-in concrete will not be subject to segregation in a dry hole, provided the bottom of the pipe is always kept below the surface of the concrete, and the tremie pipe is large enough to avoid excessive friction.

Tremie Placement Below Water Before tremied or pumped-in concrete is used, care should be taken that there is no water entering the hole; otherwise the hole (or casing) should be filled with water (or in some cases drilling fluid) to the very top. With the water level at the top, any leaks occurring will move water out of the pier, not into it. No dilution or segregation of concrete can then occur due to water entering while the hole is being filled with concrete. This is an important precaution.

When concrete is placed below water by tremie, a common source of dilution or segregation is water in the tremie pipe before concrete placement is begun. The pipe should be empty (except for air) when the pour is started. It should be resting on the bottom of the hole. This means that a closure of some kind will have to be applied to the open end of the pipe before it is inserted, it will have to stay in place until the end is on the bottom and the concrete in the pipe has reached a level where the concrete pressure exceeds the water pressure at the bottom, and it must come off readily when the filled pipe is lifted slightly. A special closure of plywood is satisfactory, and is light enough to float out on top of the rising concrete—if it does not get hung up in the reinforcing cage. If the plywood disk does not float out, it is sufficiently incompressible that it would not be expected to weaken either the footing or the shaft appreciably. Another closure that some contractors have found satisfactory is an inflated basketball. The expedient of blocking the end of the tremie pipe with rags, or with anything that would be likely to produce a serious weakness wherever it lodged, should never be condoned.

The Tremie Pipe   The authors' experience indicates that a tremie pipe should be at least 8 in. (20 cm) in diameter, and that 10 or 12 in. (25 or 30 cm) is better. Some authorities recommend that the diameter of the tremie pipe be at least eight times the size of the maximum aggregate.

Several instances of serious weakening of concrete by contamination from contact with aluminum pipe have been reported [4.15]. The use of this metal to convey fresh concrete should be avoided under all circumstances.

Advantages of Pumped Concrete   Although pumped concrete is usually more expensive than tremied, it has certain potential advantages that may prove more important than the cost differential. When it is available at all, it can usually be obtained in strengths up to 6,000 psi (420 kgf/cm^2). Where the need for it can be anticipated, the design value of f'_c can be increased. Because discharge can be at the bottom of the hole until the entire column has been concreted, continuity of the concrete without segregation can be assured. In some instances this will justify the use of a smaller concreted shaft size, and the saving in concrete volume will go far toward compensating for the increased cost per cubic yard for the pumped concrete.

Slump and Maximum Aggregate Size   Slump and maximum aggregate size must be carefully controlled in order to be sure that either free-falling or tremied concrete will flow freely between reinforcing bars and completely fill the space outside the cage. Sometimes it is advantageous to govern aggregate shape as well as size. Some authorities specify gravel rather than crushed stone for coarse aggregate in concrete to be used in reinforced piers to avoid the "harshness" associated with crushed stone mixes [4.16].

When piers with reinforcing cages are used, the slump of concrete should be 6 or even 7 in., and for tremie placement it may be more. One major contractor reports using a 7½-bag mix, with 7- to 9-in. slump, in piers where a small amount of shrinkage will do no harm. For piers without reinforcing, slump may be as low as 4 in. These items should be covered in the plans and specifications (see Chap. 7).

High slump can—and must—be combined with suitable strength. This can be accomplished by adding cement—at the expense of shrinkage—or by the use of suitable additives. It cannot be done by adding water to a well-designed mix with low slump.

Particular caution must be used in the acceptance of ready-mix deliveries in very hot weather. The concrete will take its initial set sooner, and sometimes a delivery man will add water to be sure he can discharge the load satisfactorily—with disastrous results to concrete strength. On the other hand, to go ahead and place stiffening concrete with a lower slump than the design calls for can easily result in voids and honeycombing. Hot-weather concreting calls for special care in design, delivery, and placement. It may be necessary to use a retardant in the mix.

Vibration of concrete in drilled piers is sometimes required, most often in the top 10 to 15 ft (3 to 4.5 m), and may be very helpful in assuring that the concrete makes complete contact with reinforcing bars, as well as avoiding voids in the part of the pier where hydrostatic pressure of the wet concrete is least. Down-hole vibrators suitable for use in deep piers are becoming available, and their use will probably increase. However, it has been reported that vibration of concrete inside a temporary casing will sometimes cause packing and wedging of sand and gravel outside the casing, resulting in considerable difficulty in pulling the casing; and when used in a harsh or dry, low-slump mix, it may cause the concrete to seize the inside of the casing so that it cannot be pulled without lifting the concrete.

4-5.2 Pulling the Casing

When a hole has been cased to seal out water (as distinguished from casing placed to protect personnel or to prevent minor sloughing of sidewall from contaminating the bottom), the casing must not be disturbed until enough concrete has been placed to produce a higher concrete pressure at the level of the casing seal than the water pressure outside the casing at that level. If this precaution is neglected, water will enter the concrete column when the casing seal is broken (Fig. 4-27). The result will be a weakened column, and in some cases an open void or a zone of aggregate washed clean of cement. If there are cavities of substantial volume outside the casing, concrete will flow out from the casing to fill them, and the level of concrete in the casing may drop far enough for it to no longer exceed the outside water pressure. This means that attention must be given to maintenance of concrete level in the casing as it is pulled, particularly when cavities may have been developed during the drilling. (Good drilling practice requires that such cavities be "packed off" with clay soil during drilling and before casing is placed.)

A circumstance which is especially likely to give trouble is a sealed-off

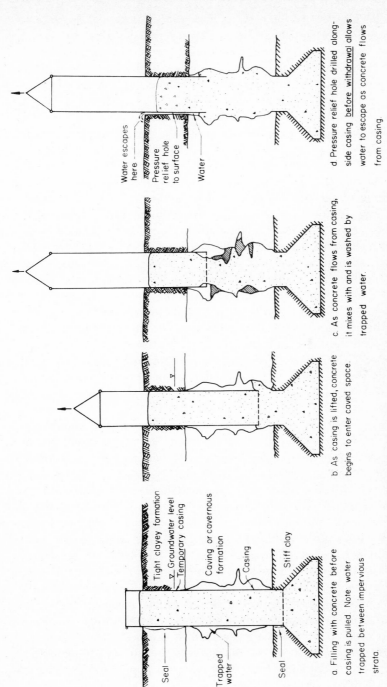

Fig. 4-27 This sketch illustrates (a,b,c) contamination of pier concrete by groundwater, and (d) avoidance of contamination by maintaining concrete pressure higher than groundwater pressure during pulling of casing.

a Filling with concrete before casing is pulled Note water trapped between impervious strata

Tight clayey formation
Groundwater level
Temporary casing
Caving or cavernous formation
Casing
Stiff clay
Seal
Trapped water
Seal

b As casing is lifted, concrete begins to enter caved space.

c As concrete flows from casing, it mixes with and is washed by trapped water.

d Pressure relief hole drilled alongside casing before withdrawal allows water to escape as concrete flows from casing

Water escapes here
Pressure relief hole to surface
Water

zone with artesian pressure above the initial bottom seal of the casing.
If an artesian zone exists and is vented to ground surface by clearance
between casing and hole, there can still be a positive head of concrete
against the entry of water into the column as the casing is raised. A
sealed-off artesian zone can be vented to the surface by drilling a small
hole to its level, alongside the casing (Fig. 4-27*d*).   Care must be taken,
if this is done, not to disturb the casing's bottom seal.

When telescoping casing is used, with the lowest section sealing off
groundwater, it is necessary to place enough concrete in the lowest
casing section to be sure that, when the casing is pulled and the bottom
seal is broken, the surface of the concrete level does not fall below
the bottom of the next larger casing (Fig. 5-6).   This will depend on
the relative positions of the top of the bottom casing, the bottom of
the next section, the level of groundwater outside the bottom section,
and the volume of water to be displaced outside the lower casing and
below the bottom of the next section.   If there has been much caving
during drilling, or if the lower formation is cavernous, the concrete
level should be watched carefully as the bottom section is lifted; and
more concrete should be added if it appears that concrete level is about
to fall below the bottom of the next section, thus admitting groundwater.   The need for this precaution is illustrated in Fig. 4-28.

Another circumstance that can lead to serious mistakes in pulling
casing is the presence of a cased-off stratum of soft "squeezing" soil,
which can produce "necking down" of the pier shaft when the casing
is removed.

When a reinforcing cage has been placed inside the seated casing
and the casing has been filled with concrete and is then pulled, forces
can be developed in the cage that it was not designed to withstand,
especially if the concrete is stiff or harsh.   In order to fill the annular
space exposed below the rising casing bottom, as well as any additional
cavities in the sidewall, the concrete inside the cage must slump and
flow downward, putting the cage in hoop tension; and the concrete
between casing and cage likewise flows downward, adding to the vertical
compression force on the cage.   When the pier hole has been drilled
through caving or cavernous ground, the slump of concrete as the casing
is pulled can be very substantial.   It is not uncommon for a spirally
wound reinforcing cage to fail at this stage by wracking, the twisting
action expanding the spiral turns while shortening the longitudinal members.   A positioned reinforcing cage, which is expected to remain stable
while the casing comes out and the concrete goes in, can shrink vertically
and disappear from sight entirely, leaving the top of the shaft unreinforced and the observers mystified (Fig. 4-29).

This phenomenon, as indicated above, may become serious when the

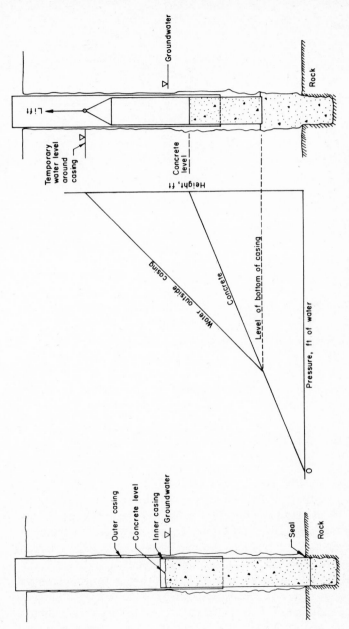

Fig. 4-28 Telescoping casing, showing relation between groundwater pressure and concrete pressure.

a. Inner casing filled with concrete, ready for pulling

b. Inner casing partially lifted; slumping concrete forces water to rise in space around casing. Concrete and water pressures are equal at interface.

153

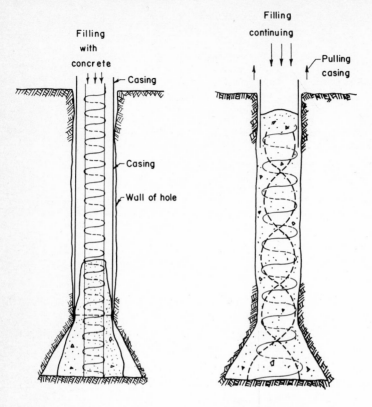

a. Concrete flowing down and out through cage puts vertical bars in compression and spiral tied bar in tension.

b. Reinforcing cage fails by wracking, with tie spiral expanding in diameter and cage shortening in length.

Fig. 4-29   Distortion of an inadequately tied reinforcing cage by weight of low-slump or harsh concrete.

space to be filled outside the casing is substantial. It can also occur as an underream is being filled with concrete through a cage resting on the bottom. It is amplified by the use of too harsh a mix, by the use of concrete with too large aggregate, by too close bar spacing, or by carelessly made or inadequately designed ties in the cage. It can be minimized by use of concrete with a slump of 6 or 7 in.

Casing should be pulled in an axial direction, never by an eccentric pull. It may be necessary to break the seal by a short jerk; otherwise the casing pull should be smooth, without jerks. Concrete level should be maintained so that it never falls below the bottom of the casing until the level of the surface casing (or design level for top of pier, whichever is lower) has been reached.

If the concrete has a suitable slump, and the casing is pulled promptly, as described above, no trouble should develop during the pulling. If the concrete is too harsh, or if the casing is jerked or pulled too fast, the concrete column can be pulled apart below the bottom of the casing, with disastrous results to the pier. When this occurs, the reinforcing cage may be found to have risen several feet by the time the casing is out of the ground.

Many contractors have found that a vibratory hammer, such as the Foster D-50, is very effective, not only for driving casing but for extracting the casing smoothly without lifting the concrete.

Special attention should be given to the ground adjacent to the casing during pulling. The development of subsidence at that time indicates a strong probability of intrusion of soil or water into the newly exposed column of concrete.

4-5.3 Interrupted Pour

Although the placement of concrete in a pier should be continuous from start to finish, this is not always possible. When a partially filled pier is allowed to stand until concrete has taken its initial set, the exposed surface requires attention before more concrete is placed, to assure good bond between the old and the new concrete. Laitance should be removed, and the surface should be roughened and slushed with a 1:1 cement grout just before the pour is resumed. Where the reinforcing cage prevents direct access to the surface, it is imperative that all laitance be removed.

4-5.4 Uncased Pier through Water

An ingenious technique has been used to produce smooth-finish concrete piers through water without the necessity of leaving costly steel casing in place and subject to corrosion [4.12]. The procedure is to drill the pier as usual, setting a smooth steel casing (not a corrugated or galvanized liner) to contain the concrete, but using a second casing, at least a foot larger in diameter than the other, centered with the pier, set far enough into the sea bottom to hold firmly, and securely braced in place. The annular space between the two casings is filled with clean sand, and concrete is placed in the inner casing. The inner casing is immediately pulled with a vibrating puller, using care to get a smooth vertical pull. The annular mass of sand retains the heavier concrete without any tendency to slough or intrude in the pier. After the concrete has set, the outer casing is pulled (loosening the sand annulus by jetting if necessary), the sand drops off, and the pier is exposed, with a smooth surface.

4-6 DEFECTIVE PIERS

4-6.1 Kinds and Causes of Defects

Defects in completed piers are probably no more common than defects in any other type of deep foundation. However, in most pier-supported structures, the column load is supported by a single pier, whereas a group of piles would be required to support a similar load. For this reason, a single defective pier is more likely to result in structural damage or failure than is a single defective pile or a proportionate percentage of piles scattered through the project. This is one of the reasons why, in Chap. 5 of this book, the necessity for thorough and competent construction supervision and inspection has been emphasized. If these functions are adequately performed, any construction defects should be avoided or detected and remedied before a pier is completed, and the completed pier should be adequate to carry the design load without suffering detrimental settlement or deflection. However, experience shows that completed piers do occasionally turn out to be defective; and there have been several instances of near-disasters, avoided more by good fortune than by care and vigilance [4.17, 4.29].

The types of defect that can occur and the conditions that can lead to these occurrences have been reviewed by Baker and Kahn [4.17]; they may be summarized as follows:

1. Inadequate concrete strength, caused by segregation during placement, by improper placement in cold joints, or by delivery of poor concrete to the site.

2. Displacement or contamination of the concrete by caving or by collapse of walls of hole or underream.

3. Voids or discontinuities in the shaft or underream, caused by use of concrete which is too stiff; by collapse of casing or liner; by squeezing in of soft formation; or by hang-up of concrete in the casing while it is being pulled.

4. Inadequate bearing, resulting from inadequate soil investigation and evaluation, from error in identification of the proper bearing material while the hole is being drilled, or even, occasionally, from an unauthorized change in dimensions of the bearing area or omission of the underream.

If investigation and design were always perfect; supervision always competent and adequate; inspection always continuous, experienced, and alert; and the contractor's personnel always expert and conscientious—if all these conditions prevailed, all the time on a project—then there would be no defective piers.

4-6.2   Checking Completed Piers

For important or settlement-sensitive structures, and when large loads are to be carried on drilled pier foundations, the geotechnical investigation, construction supervision, and inspection should be performed meticulously. Also, however, consideration should be given to the advisability of postconstruction investigation to confirm the continuity and integrity of the completed piers on such projects. Not every such project will need this kind of check. When the engineer can be assured that piers have been drilled through formations that give no trouble, and that they have been adequately inspected and filled with concrete of the correct slump from a reliable source, with no water in the holes and with the work done by experienced and conscientious contractor's personnel, he can be reasonably certain that they are sound and continuous. But if any of these conditions deviates from the ideal, and particularly if there has been trouble in getting a dry hole or if the concrete has been placed under water (tremied or pumped in), then a check on the condition of the completed work becomes advisable.

At this time, the most direct and reliable check on pier dimensions and soundness is obtained by diamond coring. The larger the diameter of the core, the more information can be obtained and the more expensive the test. For this application, the authors do not recommend the use of cores smaller than NX ($2\frac{3}{8}$ in. or 60 mm); larger cores, up to 6 in. (15 cm) diameter, have been used. The choice of core size, of the number of core holes per pier, and of the individual piers to be investigated depends on the reasons for making the test in the first place. The construction events or difficulties (if any) which led to a judgment that test core borings should be made should be evaluated to determine what defects may have resulted that should be looked for. For example, a 10-ft-diameter (3 m) pier for a very important building, in a situation where there was a strong suspicion of the presence of defective concrete, was tested by 11 diamond core borings [4.17]. A pier for moderate loads, installed under circumstances that had not led to any suspicion of its integrity but where general policy dictated testing of a selected percentage of the finished piers, might be tested by only one diamond core boring.

In any case, where diamond core borings are used to test pier concrete, the work should be done by the most skilled contractor available. His equipment should be in first-class condition, and his drillers must be experts in *diamond coring*. In the authors' opinion, this requires a negotiated contract; and a price per hour, with a provision for payment for diamonds used, is fairer to the contractor than a price per foot of core. Diamond coring in concrete is difficult work at best. When zones of poor concrete, or of washed gravel, are encountered, difficulties and uncertainties are multiplied. An unskilled driller (or a worn bit) can produce results that make conditions look much worse than they really are. The problem is too important and the results too critical to justify trying to save money on the test-boring contract.

Diamond coring can be supplemented by other test procedures that often can expand considerably the information yielded by the cores and drilling records. The borehole camera [4.34, 4.35], usable in NX and larger holes, can be used to determine the extent of cavities or of layers of uncemented material—information often not determined accurately from cores. Automatic caliper logging, seismic or sonic velocity logging, and gamma ray logging have been used to advantage. Inclinometer logs have been used to determine straightness and plumbness of test borings (conditions which are rarely determinable by direct observation) [4.17, 4.36, 4.37].

Occasionally, underreams (bells) have been omitted, either by inadvertence or as a consequence of poor supervision or inspection. This kind of defect seems improbable, but it has occurred [4.29]. If such an omission is suspected, or if there is reason to suspect that an underream may have caved or collapsed before or during concrete placement, a conventional test boring can be drilled close to the shaft. It is difficult to be assured of the plumbness of a deep test boring, and for this reason, failure to encounter the concrete of the underream may be inconclusive unless the actual alignment of the test hole is checked by suitable instrumentation (inclinometer, etc.).

4-6.3   Repair of Defective Pier

In instances where a defect is located in the upper part of a pier, either above the water table or in a situation where the water level can be drawn down readily, the best and surest method of repair is to make an excavation alongside the pier (perhaps with a pier-drilling machine). The defective material should be removed and replaced with good concrete (in some cases adding dowels or additional reinforcing to compensate for uncertainties in bond at cold joints).

Grouting is always done blindly; with the best techniques and materials, the ultimate location of the grout that has been pumped into a hole is a matter of conjecture and speculation. A grouted zone in a pier always should be checked by diamond borings and by any other instrumentation that has been used in locating the defect.

Short defective lengths of a pier can be repaired by drilling a number of large core holes down through the defective zone, placing steel rods *of sufficient cross section to carry the entire vertical load* and long enough to span the defective zone, and grouting the rods in place.

When this is done, the grouting should be planned not only to cement the rods in place, but also to fill the voids in the defective zone as far as possible. This method of correction is expensive and requires meticulous work, but for deep piers it may cost substantially less than either complete replacement or side excavation and manual repairs.

Grouting, like diamond coring, requires the best equipment and techniques. Moreover, if it is to be successful, it requires special attention to the choice of a suitable grouting material. For large open voids, an ordinary neat cement grout, or even a sand-cement grout, might be useful. If penetration through gravel or coarse sand is required, however, then a less viscous grout is needed. Cement grout with an additive to reduce its viscosity, or grout in which the cement is dispersed by a high-speed machine such as the "Colcrete" mixer [4.38], will penetrate readily through gravel and coarse sand, whereas ordinarily prepared cement grout will not. For penetration of finer sands, a chemical grout of the kind that sets up with a strength comparable to that of Portland cement concrete (such as one of the sodium silicate grouts) should be used.

Occasionally, the least expensive way to repair (or replace) a defective pier will be to construct new drilled piers on opposite sides of the defective one, picking up the column load on a short girder bridging over the defective unit. This technique is usually possible only for correcting defects discovered before the superstructure has been erected. Correction at a later stage will usually involve conventional—and very expensive—underpinning operations.

4-7 HOUSEKEEPING—JOB ORGANIZATION

A good, smooth-running drilled pier job requires more attention to some surface details than does, for example, a pile foundation job of the same size.

Surface drainage is especially important in pier-drilling operations. Drainage should always be away from a hole being drilled. The surface casing (if any) should be high enough to prevent fall-in of spoil or drainage of surface water into the hole, and a slight mounding of spoil around the hole may be allowed to keep drainage away from the hole until concrete has been placed. Good surface drainage over the entire site will facilitate the work in bad weather, especially where clayey soils are involved, and care should be taken to prevent ponding whenever the job is shut down.

Spoil from the drilling must be disposed of. Clayey spoil spilled in the work area may seriously impede operations; and if carelessly spilled on public thoroughfares while being hauled away, it can even cause a job to be shut down. Loose spoil left too close to an open hole may fall in, impeding the work, endangering workmen in the hole, or contaminating concrete when the casing is pulled or during or after concreting.

Open holes must be covered securely whenever the job is shut down, not only as a safety measure, but also to prevent entry of rain or snow. An open hole should *never* be left uncovered overnight.

Unlike most piles, drilled piers require careful scheduling of deliveries. In most circumstances, concrete placement must follow promptly on completion of drilling. This requires precision timing of concrete delivery, of delivery and placement of reinforcing cages, and of cranes for handling cages and for placing and pulling casing.

4-8 DRILLED PIERS FOR RETAINING WALLS

The application of drilled pier techniques is becoming increasingly common in the construction of retaining walls for excavations [4.19–4.21, 4.25, 4.30, 4.31]. An outstanding advantage of this type of retaining wall is that it can be designed to be watertight and to take full earth and groundwater pressure. Such walls can, under many circumstances, be designed with tiebacks, thus requiring no bracing within the excavation, and therefore offering no impediment to construction operations within the excavation.

This type of construction is often done in a sequence as follows:

1. A series of pier holes are drilled around the periphery of the proposed excavation, to a depth well below the proposed bottom of excavation and spaced less than two diameters apart, center to center. Care must be taken in drilling these holes, so that an open hole, or concrete that has been placed but has not yet set, is not disturbed by the drilling of an adjacent hole. It is often necessary to drill and concrete alternate holes first, then return and complete the series after the concrete of the first group has set.

These holes are concreted and reinforced as piers, designed to withstand external soil and groundwater pressure on the excavation wall when the excavation is complete. The design may involve taking the side pressure either by cantilever action, as shown in Fig. 4-30a, or by the installation of posttensioned earth or rock anchors, as illustrated in Fig. 4-30b. In some designs, some or all of these piers are also used as the outer foundation piers for the structure. The foundation piers will usually have to be drilled deeper than the others to provide adequate bearing for the structural loads, in addition to whatever lateral resistance is needed for retaining-wall action.

2. When the concrete in the first series of piers has set up, a second

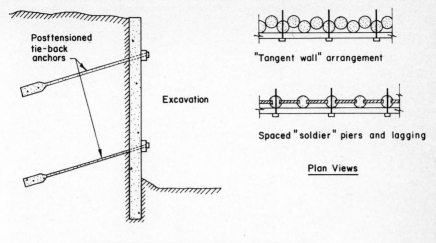

"Tangent wall" arrangement

Spaced "soldier" piers and lagging

Plan Views

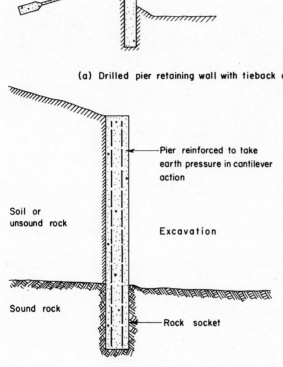

(a) Drilled pier retaining wall with tieback anchors

(b) Cantilever-type drilled pier retaining wall

Fig. 4-30 Types of drilled pier retaining wall.

series of holes is drilled, intermediate between the piers of the first series and displaced enough toward the outside of the excavation that each hole touches the concrete of the two adjacent piers (hence the name "tangent wall" that is sometimes used). These holes are cleaned carefully to make sure the concrete of the adjacent piers is exposed; then they are filled with concrete, completing the wall. The second

Fig. 4-31 Drilled pier retaining wall with two levels of posttensioned tiebacks (Municipal Parking Garage, Dallas). (*Mason-Johnston Associates, Inc., Dallas, Tex.*)

series of piers may go to the full depth of the first series or may be stopped shortly below basement floor level, depending on soil or rock conditions. They may be reinforced to take some of the thrust of soil and groundwater pressures or not, depending on the design and reinforcing of the first series.

3. When the wall has been completed as described above, the excavation is started. If lateral earth pressures are to be taken by an earth anchor system instead of by cantilever action in the piers, excavation is interrupted as soon as it is deep enough to permit drilling of the anchor holes. The anchor holes are drilled and anchors installed, the anchor rods are posttensioned, and excavation is resumed down to the next level of anchors (if any) [4.20]. Anchor holes are usually drilled with a special machine, using a long continuous-flight auger. They may be horizontal or sloping, depending on the soil and rock formations to be penetrated. When rock anchorage cannot be obtained, sometimes

conventional pier holes with underreams are put in, at a suitable batter, with pier-drilling machines.   Figure 4-31 shows a drilled pier retaining wall with two levels of posttensioned anchors.

4. In many cases, the drilled pier retaining wall is combined with the basement wall by casting a concrete facing against the inside of the retaining wall.   In other cases, a separate wall is constructed, the space between the piers and the inner wall being used as a drainage gallery to take care of seepage through the outside wall.

Many variations of these designs have been worked out to fit the requirements of the structure and the local soil or rock conditions. Sometimes the spacing between the holes of the first series is nearly two diameters, and the holes of the second series are the same diameter as the first; or the first series may be closer together, with the second series much smaller.   Sometimes only one series of piers is installed, with a spacing of three or more diameters; grooves for insertion of precut lagging are formed by positioning metal forms, or foamed plastic strips, on the sides of the pier hole before concrete is placed; and lagging is added as the excavation progresses [4.19].

Fig. 4-32 A "tieback" drill for anchor holes. (*Watson Manufacturing Company, Fort Worth, Tex.*)

Figure 4-32 shows a specialized drilling machine for drilling small-diameter anchor holes in soil or rock.

Drilled piers have also been used in conjunction with mud slurry to construct retaining walls. The piers are drilled at a spacing of three to five diameters. The soil between piers is then excavated, the removed material being replaced by slurry as excavation proceeds. Generally, short trench sections are completed alternately; reinforcing is placed according to the design; and the slurry is displaced by concrete, pumped in at the bottom of the trench. When the concrete in the first set of trench sections has set adequately, the intervening trench sections are excavated and the wall is completed. Tiebacks are generally used to take lateral loads [4.39].

ANNOTATED REFERENCES

4.1   Drilling Experts Develop Tools to Speed Caisson Work, *CM&E,** January 1963, pp. 80–83 (reprint from Case International, Chicago).

Tools developed by Case are illustrated: core barrel for rock, boulder rooter, belling bucket, mucking bucket with rubber flap valves, chipping chisel, and a special beam for pulling casing.

4.2   Glidden, H. K., Floating Rigs Build Piers, *Roads Streets,* February 1965 (5 pages—reprint from Hughes Tool Co., Houston).

Forty-four pier shafts for bridge across Sandusky Bay, Ohio. Holes 7 ft in diameter, average 60 ft below water and 10 ft into sound limestone. Bid cost $305 per cu yd of concrete below water level. Special Williams (Hughes) rig, 88-in. roller-bit cutting head, reverse circulation. Bit was weighted to 30 tons by adding hollow segments, filled with steel punchings, on top of cutter head.

4.3   Big Bit of Drilling Required for St. Louis Medical Center, *Construct. Dig.,* Nov. 7, 1968.

One hundred eight piers, drilled with Williams (Hughes) CLLDH on Manitowoc 3900B. Holes were 36, 42, and 48 in. diameter, into hard dolomitic limestone. Bit was multiroller-type, air-cooled and cuttings blown out by air, and weighted to 55,000 lb by lead filling in skirt of bit. Contractor: Drilling Service Co. of St. Louis. Bells were made by hand labor and air tools.

4.4   Gauntt, C. G., Marina City—Foundations, *Civil Eng.,* December 1962, pp. 61–63.

Sixty-six-in.-diameter reinforced caissons for towers of Chicago's Marina City, installed by "Case" method, to average depth 110 ft, through bouldery glacial till to bear on sound limestone. Groundwater sealed off by temporary casings; thin steel liners used for forms for shafts, and casings were salvaged and reused.

4.5   Rigid Casings Speed Wet Sand Bores, *CM&E,* March 1964 (4 pages—reprint from Case International, Chicago).

Thirty-three piers to support American Dental Association Headquarters Building in Chicago. Nine piers were 8 ft diameter, with underreams to 24

* *CM&E = Construction Methods and Equipment*

ft diameter, for column load of 8,700,000 lb. Borings went through saturated sand, soft and stiff clay, by patented "Case" method, forming slurry of sand and bentonite through 30-ft sand layer; 10-ft-diameter casing set through slurry-filled hole and sealed into clay by forcing down several feet; then slurry was removed and 8-ft-diameter hole drilled neat through clay, in the dry. Smaller steel liner was set temporarily through clay while 24-ft bell was mined by hand, using pneumatic spades and paving breakers.

4.6 Biggest Mechanically Built Caisson, *CM&E* (2 pages—reprint from Case International, Chicago).
Describes same project and procedures as Ref. 4.5.

4.7 Mohan, D., G. S. Jain, and D. Sharma, Bearing Capacity of Multiple Underreamed Bored Piles, *Proc. 3rd Asian Reg. Conf. SM & FE, Haifa*, pp. 103–106, 1967.
Report of experiments indicating that load settlement of underreamed piers in sandy and clayey soils is substantially improved if two or more underreamed bulbs are provided, and that an increase in bearing capacity of about 50 percent can be obtained with only 30 to 40 percent of the concrete required for a straight-walled pier of the same diameter.

4.8 Moore, W. W., Foundation Design—the Golden Gateway, *Civil Eng.*, January 1964, pp. 33–35.
Heavy structures in San Francisco's Golden Gateway project are founded on skin-friction piers in weathered sandstone and shale. Piers pass through variable depths of fill and bay mud. Design stresses of 10,000 psf were used, based on a load test in which support was by sidewall shear only. Holes were drilled 54 in. diameter in soil, 48 in. diameter in rock, with casing sealed into rock. Drilling was done in mud slurry made from soil being drilled; slurry was thinned to "consistency of heavy cream or buttermilk"; density was held to not more than 85 pcf. Smear on rock sockets was cleaned by scraper teeth on outside of cleanout bucket, sometimes by water jet.

4.9 Drill Rotates Caisson Liners into Bedrock, *CM&E*, August 1964, pp. 96–99.
Pier foundations for a Chicago 30-story building were drilled through clay, hardpan, silt, and loose rock to rest on bedrock at about 100-ft depth. Bottom liners of fabricated steel were twisted into the rock to form a seal for dewatering.

4.10 Big Drills Twist 100-ft Caissons into Place, *CM&E*, May 1969, pp. 62–66.
Seventy-nine casings up to 11.5 ft diameter were sealed into rock or hardpan at depths of 80 to 112 ft for the 52-story IBM building in Chicago. Casings were left in place as required by current Chicago building code. Drill was Calweld CH on Vicon 3900 Manitowoc. Drill develops 400,000 lb-ft of torque. Casing barrels weigh 15 to 44 tons each. To limit pressure on cutting teeth, the casing is suspended from a second kelly. Two to four ft of penetration into rock is sufficient to seal off water. Manitowoc had to be reinforced, counterweight added, and lagging and sheaves changed to handle the cable. Contractor was Caisson Corporation, Chicago.

4.11 Caissons Twisted into Rock under Water Carry Elevated Highway over River, *CM&E*, April 1968 (6 pages—reprint from Hughes Tool Co., Houston).
Three bridge piers for Interstate 435 over Missouri River near Kansas City, each with eleven 6-ft steel "caissons," the concrete and rebars extending 15 ft into rock. Eight-ft-diameter casing is twisted down by Hughes rig capable of exerting 1,000,000 lb-ft of torque; sand is then drilled out in

6.25-ft hole down to rock, adding clay and water to make a slurry of the sand (patented "Case" method). Six-ft-diameter casing is then inserted, twisted 3 ft into rock, and emptied of slurry. Then 5.5-ft auger hole is drilled 15 ft into shale rock, and hole is cleaned out manually. Outer (temporary) casing is pulled after concrete has set. Contractor used Hughes CLLDH rig and tools from Case International, Chicago.

4.12   Smith, M. L., Sheath of Sand Salvages Steel Caisson Liners, *CM&E*, October 1964, pp. 118–123.

An outer (temporary) casing is set through water and sand and sealed into clay, an inner temporary casing is centered in the first one, and the annular space is filled with sand. The boring is then extended into the clay, the underream is formed, and steel and concrete are placed in the usual way. As soon as the concrete is in place, the inner casing is pulled carefully, leaving the concrete retained by a smooth layer of sand. When the concrete has set, the outer casing is pulled and the sand sloughs off, leaving the smooth column of concrete. Illustrated.

4.13   Bad Caissons on 500 North Michigan Avenue, *ENR*, Sept. 29, 1966, p. 14.

Trouble in drilled piers for 25-story office building is attributed to fog in the hole during cold-weather concreting, which prevented proper inspection; also, no casing was used.

4.14   "Investigation of the Concrete Free Fall Method of Placing High Strength Concrete in Deep Caisson Foundation," report prepared by Soil Testing Services, Inc., for Case Foundation Co., April 1960.

4.15   Concrete Trips on Aluminum Pipe, *ENR*, Dec. 18, 1969, p. 57.

Aluminum pipe declared unsuitable for tremie or pumped concrete; BPR has advised field offices and state highway departments to this effect. Tests produced hydrogen foaming; concrete expanded; loss of strength at 14 days was 9 to 14 percent.

4.16   Randall, Frank, Design of Concrete Foundation Piers, *Portland Cement Assoc. Concrete Rept.* XS 6830.

This paper recommends low-slump gap-graded concrete for deep piers. This recommendation is acceptable for straight-shaft piers without reinforcing cages or large underreams. It is wrong if the pier has either of these features.

4.17   Baker, C. N., and F. Kahn, Caisson Construction Problems and Methods of Correction, *ASCE Preprint* 1030, October 1969.

A good paper, going into some detail about types of flaws, the conditions which can cause them, and corrective measures.

4.18   Drilled Caissons Cut Underpinning Cost, *ENR*, Sept. 3, 1964, pp. 36 and 41.

A description and drawing illustrating underpinning of an existing building while a 20-ft basement excavation was being made adjacent to it. Grade beams spanned from rock located under the building to pairs of 18-in.-diameter piers drilled to deeper rock at the location of the proposed basement wall.

4.19   Subway Caisson Walls Trim Time, *CM&E*, September 1970, pp. 84–86.

Wall constructed by piers 3 ft diameter, casing 42 in. diameter; rebar cages were fabricated with slot for Styrofoam filler to be removed after concrete hardened, making key for 8-in.-thick, 2-ft-long diaphragm wall section. Every fourth pier had 5-ft 30° bell at bottom. Sonotube was used to position and center cage and was pulled when concrete was placed. Shaft at wall level was 3 ft diameter, 42 in. in foundation below wall. Drilled with Calweld rig on Lima 30-T crawler crane.

4.20 White, Robert E., Pretest Tiebacks and Drilled-in Caissons, *Civil Eng.*, April 1963 (3 pages—reprint from Spencer, White & Prentis, New York).
Retaining wall for an excavation formed by vertical drilled-in caissons at wide spacing, held by 45° tiebacks drilled and grouted into rock and then posttensioned (Freyssinet system). Tension in tiebacks is 150 kips.

4.21 Engstrom, H., and R. Oates, Drilled Pier Cofferdam for a Building, *Civil Eng.*, April 1967, pp. 44–45.
A three-story basement below groundwater, in an area where drawdown could not be allowed, was constructed inside a watertight cofferdam constructed of "tangent" drilled piers. One hundred thirty-four holes 30 in. in diameter were drilled at 3-ft centers and concreted into rock and held at the top by shoring, then intermediate 15-in.-diameter piers were drilled outside and tangent to the first series. A 2-ft-thick concrete floor slab was tied down with anchors drilled into bedrock to resist hydrostatic uplift.

4.22 Houston's Distinctive Underpass Is Built in and on the Ground; Dug Out, *ENR*, July 19, 1962.
Retaining walls are constructed below the water table as closely spaced drilled piers, then excavation is made, and finally a concrete curtain wall is constructed in front of the pier wall and anchored to the piers. Sand wells behind the wall are used to relieve hydrostatic pressures.

4.23 Big Down-the-hole Rig Drills Caisson Shafts, *CM&E*, December 1963, pp. 88–89 (the Magnum drill).
Description of a pier drill with seven pneumatically operated hammer drills mounted on a rotary drill assembly, operated by a crane-mounted drilling machine. The drill assembly comes in 15, 18, 24 and 30 in. diameter; it is claimed to make 15 to 30 ft of penetration per hour. In a St. Louis job, a 24-in. unit drilled 58 piers 15 ft into limestone in 13 eight-hr days.

4.24 Long-buried Tiebacks Leave Cofferdam Unobstructed, *CM&E*, May 1966, pp. 130–135.

4.25 Andrews, G. H., and J. A. Klasell, Cylinder Pile Retaining Wall, *Natl. Res. Council Publ.* 1240, 1964.

4.26 Downhole Float Controls Tools Coring Deep Caisson Shafts, *CM&E*, January 1966, pp. 67–71 (6 pages—reprint from Case International, Rochelle, Ill.).
A detailed description of techniques, tools, and sequence of operations on the foundations of the 100-story John Hancock Building, Chicago. Deepest pier, 191 ft; most to about 150 ft; diameter 10 ft; for foundation pressures of 100 tsf on hard limestone.

4.27 Tight Site for High Rise Foundations, *Construct. Dig.*, April 1966 (2 pages—reprint from Calweld, Santa Fe Springs, Calif.).
Hospital Service Corporation's 17-story building in Chicago's Loop district required 43 piers, 30 in. to 6 ft 9 in. diameter, average depth 110 ft. Drill was Calweld 200 CH, rated at 360,000 lb-ft torque, mounted on Manitowoc 3900 Vicon crane. Caisson Corporation, Chicago, was foundation contractor.

4.28 Big Augers Go Deep So Buildings Can Go High, *ENR*, May 25, 1961, p. 32.
Construction methods for 66-in.-diameter piers for 60-story apartment buildings, Chicago. Piers were 66 in. diameter, to 110 ft depth; lower 60 ft cased. Case Foundation Co., foundation contractor.

4.29 Osterberg, Jorj O., Drilled Caissons, Design, Installation, Application, *Soil Mech. Lecture Ser., Found. Eng.*, Northwestern University, January–May 1968, pp. 151–208.

4.30 Bulley, W. A., "Cylinder Pile Retaining Wall Construction—Seattle Freeway," *Roads and Streets Conf., Seattle, Wash.*, Jan. 27, 1965, 11 pp.
2,630 unusual cylinder piles (drilled piers) were used in 4 miles of retaining wall, part of which was used to stabilize a moving landslide area. Hillside cuts were as much as 100 ft deep, in unstable and potentially unstable soil. Holes of 5.5, 8.3 and 10 ft diameter (1.5, 2.5, and 3 m), were drilled to depths of 33 to 120 ft (10 to 36 m) below ground. No anchors were used. The piles acted as cantilever beams, transferring the load to stable soil (dense glacial till). Welded I-beams were used as core steel, the largest having webs 100 in. (2.54 m) in depth and flanges as large as 30 in. $\times$ 3.5 in. (76 cm $\times$ 8.9 cm). (See also Ref. 4.31 below).

4.31 Andrews, G. H., L. R. Squier, and J. A. Klasell, Cylinder Pile Retaining Walls, *ASCE Preprint* 295, January 1966, 46 pp.
A detailed discussion of the Seattle Freeway retaining wall: soil and geologic conditions, design, construction, and performance. (See also Ref. 4.30 above).

4.32 "REVERT, a self-destroying fluid additive for use in wells and test holes," information bulletin from Johnson Division, Universal Oil Products Company, St. Paul, Minn.; also several reprints of articles on "REVERT" from the Johnson Drillers Journal (same source).

4.33 New Hospital Doubles Up Over Older One, *ENR*, Nov. 13, 1969.
Foundations for a 13-story tower straddling the existing Kaiser Foundation Health Plan Hospital in Oakland, Calif., were drilled piers, 38 in. in diameter and 120 ft deep. The lower 50 ft of each pier hole was grooved in 1-ft rings on 3.5-ft centers to assure development of sidewall shear in the foundation soil.

4.34 Trantina, J. A., and L. S. Cluff, *NX* Bore Hole Camera, *Symp. Soil Exploration*, ASTM STP 382, 1963, pp. 108–120.

4.35 The *NX* Bore Hole Camera, *ENR*, June 25, 1953.

4.36 Wilson, S. D., The Use of Slope Measuring Device to Determine Movements in Earth Masses, *Symp. Field Testing Soils*, ASTM STP 322, 1962, pp. 187–198.

4.37 Digitilt, *Indicator*, vol. 1, no. 2-3, pp. 6–7.

4.38 Champion, S., and L. Truman Davies, "Grouted Concrete Construction," *Reinforced Concrete Assoc. Meeting, London*, Feb. 12, 1958.
This paper deals with grouted concrete made by the "Colcrete" process.

4.39 Kapp, M. S., Slurry Trench Construction for Basement Wall of World Trade Center, *Civil Eng.*, April 1969, pp. 36–40.

4.40 Custom-built Reverse Circulation Drill Rig Used in Constructing Offshore Platform, *Western Construct.*, April 1964, 2 pp.

4.41 Big Bell-bottom Caissons Will Support Detroit's Second Civic Center Building, *ENR*, Dec. 6, 1951, pp. 44–46.

4.42 Detroit Builds First Unit of Civic Center, *ENR*, Sept. 30, 1948, pp. 68–69.

4.43 Giant Augers Sink 15-ft-dia Missile Shafts, *CM&E*, May 1963, 3 pp.

4.44 Etheridge, D. C., Drill Digs Deep for Bridge Piers through Dam Abutment, *CM&E*, May 1970.

4.45 Caisson Trouble Hits Chicago, *ENR*, Sept. 29, 1966.

4.46 Scraper Teams and Giant Augers Pace Construction at Wyoming Missile Base Project, *Western Construct.*, July 1963, 4 pp.

4.47 St. Louis Floodwall Doweled to Rock, *ENR*, Apr. 9, 1964, 2 pp.

4.48 Vertical Mole Scores Bullseye in Deep Shaft, *ENR*, Feb. 27, 1969, pp. 28–29.

4.49 Five Wells to Drain Wet Foundation, *Western Construct.*, March 1957, 2 pp.

CHAPTER FIVE

Engineering Construction Review and Inspection

Engineering construction review and inspection practices vary widely by both organization and geographical area. In the United States legal requirements for inspection differ from city to city. Government regulations vary between various departments and bureaus. Customs have developed differently in different areas, and variation in geologic conditions has led to major differences in regional engineering construction review and inspection practices.

In this chapter the authors present their conception of a comprehensive engineering construction review and inspection program for drilled pier construction. This program represents what we believe should be followed, rather than what the present practices actually may be.

5-1 RESPONSIBILITIES OF THE ENGINEER AND THE CONTRACTOR

It should be clearly understood that the phrase "engineering construction review and inspection" applies only to obtaining compliance with the "real" intent of the plans and specifications. This implies that every reasonable effort should be made to ensure that the materials used—and construction procedures, to the extent that they are spelled out or would adversely affect the intended quality of the end product—should comply with the plans and specifications.

Engineering construction review and inspection is not intended to imply that any direction is given to the contractor or subcontractors, or any of their employees, as to how they carry out their operations. The contractor's (and subcontractors') operations are strictly their own responsibility. This includes any provisions they do or do not undertake with respect to the safety of personnel on the job site. The contractor should be given every opportunity to exercise his ingenuity in developing methods that most efficiently will accomplish the requirements of the plans and specifications.

The engineers (and architects) have the responsibility of carefully and accurately developing plans and specifications that will provide the owner with a functional structure compatible with his financial capacity, and that also will protect the safety of the public. For structures involving drilled piers, the plans and specifications definitely should include a copy of the complete geotechnical report. As previously indicated, these documents normally leave field procedures to the ingenuity of the contractor when drilled piers are indicated; however, certain procedures may be specified in order to ensure proper quality of the end product. An obvious example of the above procedure is the requirement for a dewatering system. The use of tremie concrete also might be specified under certain circumstances.

5-2 SCOPE AND LIMITATIONS OF CONSTRUCTION REVIEW AND INSPECTION

Members of the design profession should perform some on-site review on all construction projects to ascertain that the construction is proceeding, in general, in conformance with the intent of the plans and specifications. In the authors' opinion it is essential to the owner's interest to require construction review and inspection on all drilled pier installations. The extent of the services should be governed by the importance of the construction, by the complexity of the subsurface conditions at the site, and by local code requirements. In areas of known uniform geologic conditions, generally favorable for drilled pier construction, observation of the drilling operations and inspection of the completed pier holes before concreting may be adequate if light structures are involved. Even under such favorable circumstances, however, the person performing the inspection should be knowledgeable with respect to possible problems. The underground is never completely predictable, and the construction review and inspection should never be superficial or casual.

Continuous construction review and inspection are imperative on projects where the owner's investment is large, where heavy loads or special structural requirements are involved, where geologic conditions are complex, or where other unfavorable conditions could be encountered.

Where the owner and the contractor (or the designer and the contractor) are one and the same, an employee of the organization performing construction review can be subjected to pressures that could lead to an unsound decision in order to save money in construction. A man not subject to such pressures is in a position to make sounder decisions. The authors feel that it is generally to the owner's interest to have construction review and inspection done by an independent agency.

5-2.1   Definitions

Unfortunately, in the past the terms "construction review," "supervision," and "inspection" often have not been clearly defined in specifications, contracts, or other documents. In such cases there often is a tacit assumption that the meanings of these terms are clear and unarguable. The record of many lawsuits testifies to the fact that this is not true. Several decisions have been rendered by the courts holding that "supervision" in particular extends to responsibility for the adequacy. of temporary bracing or scaffolding, and even to the manner in which contractors' employees operate equipment. The use of the term "supervision" to designate the service described above as "construction review" should be avoided. "Construction review" and "inspection" require careful definition in the contract documents, with clear statements of both scope and limitations. Moreover, the definition must be tailored to the task at hand.

In recent years professional societies representing architects and engineers have made rapid strides in developing contract documents which more clearly define "construction review" and "inspection." These documents are not established to the degree, however, that their definitions will automatically be accepted in a court of law. Also, it is difficult, in such standard documents, to describe accurately such activities when drilled piers are involved. Therefore, these activities should be defined carefully in the contract documents for drilled pier projects.

The terms "construction review" and "inspection," as used in this book, refer to the activities and responsibilities of the individual, agency, or firm who represents the owner's interest in determining that the materials used and the work done are in accordance with the intent of the plans and specifications [5.1]. The engineer assigned to this duty, covering the foundation phase of a project, is often referred to as the "geotechnical engineer" or the "field engineer." "Inspection" is a part of his responsibility, and he may have project inspectors reporting to him and supervised by him.

On large projects, the owner is often represented by a "resident engineer," who remains permanently on the project and manages it.

5-2.2   Obligations

The engineer performing construction review has a direct contractual obligation to his client, who normally (and ideally) is the owner, but who may be another engineer, an architect, or an engineer-constructor. Also, he has an implied obligation to the public, as a consequence of his registration or license as an engineer and his approval of the plans and specifications. The engineer should cooperate with the contractor to avoid any unnecessary delays. However, such cooperation must not conflict with his obligations as described above.

In foundation work, or in any construction work on, in, or of soil or rock, the occurrence of so-called "changed conditions," which more properly should be termed *unexpected* conditions, is not unusual. The duty of the field engineer under such circumstances normally requires him to make field decisions. These may require variations in construction operations or techniques, materials, or even structural elements, which were not anticipated in the contract documents but which are necessary to carry out the *intent* of these documents.

The field engineer's primary obligation is to see that the work conforms with the intent of the plans and specifications; however, he also has an obligation to use his best efforts to keep the job moving smoothly, to avoid any interference with the construction schedule. During the work he provides advice and recommendations to the owner or his representative. If he finds defects in construction or materials, it is his duty to notify the owner's representative so that the error may be corrected; and as part of that duty, he keeps the contractor apprised of his advisories to the owner. A cooperative relationship between the contractor and the field engineer is always desirable, but the engineer's primary responsibility is to his client.

When the work is complete, the engineer renders a professional opinion as to whether or not it conforms to the full intent of plans and specifications.

In many instances, the field engineer will be required to submit a report to the city building inspection department (or other authority) stating his opinion regarding the conformance of the completed work to the plans and specifications. Such a report implies a judgment of the suitability of the design for the proposed use, a function that may not actually be included in the duties of the field engineer. The owner, however, will expect a final report that will be accepted by the authorities as the basis for issuing a certificate of occupancy. Thus, it is imperative that the field engineer should satisfy himself in advance regarding the adequacy of the plans and specifications, as well as what opinion or certificate he will be asked to sign. His construction review and inspection then should be thorough enough to enable him to provide a report that will meet these requirements.

As construction progresses, the field engineer should immediately advise the owner's representative of the use of any procedures or materials that do not comply with the intent of the plans and specifications. If corrective action is not taken, he then must advise the owner's representative—in writing—that he will not be able to certify that the completed work complies with the contract documents. If corrective action still is not taken, the engineer should sever his connection with the project, immediately and in writing. This action is necessary to protect both the engineer and the owner. Although the owner may not approve this action, it does protect him, not only as to the adequacy of his structure, but more importantly with respect to his obligation to protect the safety of the public.

5-2.3 Limitations of Construction Review and Inspection

It has been emphasized that the terms "construction review" and "inspection" do not imply the exercise of any supervisory authority over the contractor, including his employees, or any responsibility for their actions or lack of action. It is the responsibility of the contractor—not the field engineer—to run the job. The field engineer should not direct the contractor in the performance of his work, nor ask him to make changes outside the scope of the contract, unless the owner's representative has issued him written instructions to do so. Even though—in his judgment—the field engineer sees work being performed incorrectly, he must resist the temptation to "take charge" and direct the workmen to correct it. To yield to such temptation—no matter how obvious the error is, or how easily corrected—would put the field engineer in the position of sharing with the contractor the responsibility for the performance of the work. In the event of a dispute, the record of such an action would permit the contractor to claim that he had relied on the advice of the field engineer in the carrying out of his (the contractor's) operations throughout the rest of the job. If changes become necessary or desirable, the field engineer can then, under the conditions described above, make *recommendations* for corrective action. Any *directives* to the contractor should come from the owner.

When drilled piers are involved on a project, the field engineer frequently may find it necessary to request changes in operational techniques or in the quality or quantity of materials being used, or even to request additional work outside the scope of the original contract. Therefore it is imperative that the scope of his authority with respect to construction review and inspection be clearly described in all contracts, documents, or agreements that could affect the engineer (or architect) involved. Such documents include the contract between the owner and the contractor. The engineer (or architect) performing construction review should require, in this document, protection with respect to "injured workman" occurrences (discussed in Chap. 8).

5-3 DUTIES AND QUALIFICATIONS OF GEOTECHNICAL ENGINEER AND FOUNDATION INSPECTOR

Because of the specialized nature of the problems and the frequent need for continuous expert attention, engineering construction review involving drilled piers is often delegated to a specialist (person or firm), generally designated as the "soil engineer" or "soil and foundation engineer." The authors will refer to such a specialist as the "geotechnical engineer," a term that embraces the fields of soil and rock mechanics, groundwater hydrology, and engineering geology as they apply to foundation and earthwork problems. The duties of a geotechnical engineer, in carrying out construction review (and inspection) with respect to the geophysical phases of a project, include the following functions (which must be at the request of the owner's representative, who frequently is the architect on building projects involving drilled piers):

1. Observation of construction procedures, noting deviations from the true intent of the plans and specifications
2. Advising and conferring with the owner, architect, structural engineer, and contractor
3. Preparing specification revisions when needed and requested by the owner's representative
4. Acting as "interpreter" to be sure that owner, architect, structural engineer, and contractor understand each other in geotechnical matters
5. Responsibility for foundation construction review, inspection, and reporting

The *inspection* duties involved in construction review which are discussed here may be performed by a geotechnical engineer; or they may be performed by a technician-inspector who reports to the owner, to the project engineer or architect, or to a geotechnical engineer. In the discussion that follows, the term "foundation inspector" will be used to designate the person performing this function, regardless of whether he is a technician or an engineer. Figure 5-1 shows a typical arrangement of lines of authority, from owner down to inspector.

The foundation inspector on a drilled pier operation is required to make field decisions on matters much more complex (and less certain) than an inspector on typical concrete or steel construction, for example. If the inspector is a technician (or an engineer who is not registered or licensed), he acts as the "eyes" of the geotechnical engineer, whose duties include the exercise of engineering judgment. The inspector observes, makes reports, applies criteria established for him by the geotechnical engineer, and limits his field decisions to the application of these criteria. He refers matters requiring engineering decisions to the geotechnical engineer. But because the need for field decisions of an engineering grade is common in drilled pier work, lines of communication between the foundation inspector and the geotechnical engineer must be well established, and the foundation inspector must be able to get prompt decisions whenever questions requiring engineering judgment arise. Delay in obtaining such decisions risks impeding the contractor's operations. The making of such decisions by personnel not technically or legally qualified could involve the responsible agencies in dangerous and unnecessary liabilities.

It will be assumed, in the discussion that follows, that the person acting as foundation inspector will have the training and ability to perform the duties described. A foundation inspector is qualified by training and experience, not by appointment or self-proclamation. The degree of training and experience required depends on the size and importance of the structure and its relation to public safety, as well as on the complexity of subsurface conditions. A man may be quite competent to act as inspector of drilled pier construction for a one-story warehouse located on relatively good ground, but lack the training needed to perform the same functions for pier holes in weathered or decayed rock or for a bridge or a high-rise apartment building.

Some jurisdictions register or license inspectors (for example, San Diego, California). The organization supplying or engaging foundation inspectors should make sure of compliance with all relevant laws and ordinances.

For a large drilled pier project that requires continuous geotechnical engineering construction review, the usual arrangement is to use foundation inspectors who report to the geotechnical engineer (see Fig. 5-1). Their duties usually are (1) to make observations and records and report to the geotechnical engineer for decisions, (2) to identify bearing strata, and (3) to observe the contractor's operations for compliance with plans

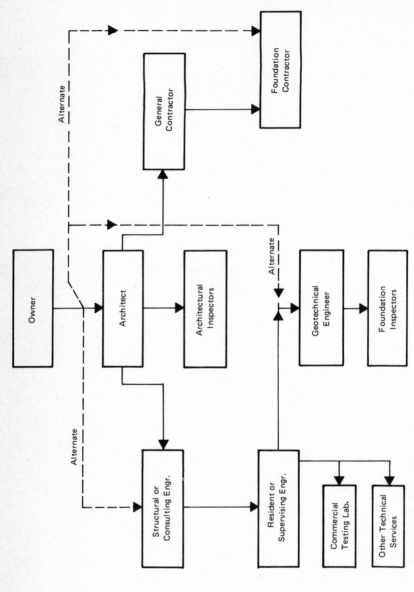

Fig. 5-1 Lines of authority in the field organization of a typical building construction job.

and specifications and notify the geotechnical engineer of any deviations. In this case, field decisions will be made by the geotechnical engineer, or, in his absence, by the foundation inspector, acting on the basis of specific criteria and instructions given him by the geotechnical engineer. It is obvious that this action by an inspector requires a special degree of technical competence. The authors have seen many instances where drilled pier costs ran far over contract or budget costs because the "inspectors" did not recognize suitable bearing material and did not approve stopping the drilling when competent soil had been reached.

Some construction contracts state that the field engineer or the inspector has the authority to require the contractor to comply with all applicable safety standards and codes. Neither the geotechnical engineer nor the foundation inspector should assume this duty unless directed to do so in writing, for to do so is to incur (or share) a liability that properly belongs elsewhere. However, if the construction contract names the geotechnical engineer (or soil engineer, or soil and foundation engineer, or even "the inspector" without designating specifically the foundation inspector), the geotechnical engineer would be well advised to obtain written instructions covering this matter. With this mention in the construction contract, and lacking specific instructions *not* to assume this duty, he should assume that a judge or jury would interpret the construction contract as having charged him with responsibility for the contractor's observance of safety rules—and that he will therefore share the contractor's liability in the event of a construction accident, even though his own contract or instructions may not mention the subject [5.2]. In this case he is incurring a liability that he is probably not being adequately compensated for—a condition that he should not accept [5.3].

In the author's opinion, the obligations and qualifications that go with specialist positions for construction review and inspection are as follows:

Geotechnical Engineer

Obligations The general duties and responsibilities of the geotechnical engineer with respect to his assignments are to represent the interests of the owner (and indirectly of the public), and to use his knowledge and expertise, acting in an *advisory* capacity, to keep the owner apprised of any construction errors which could lead to increased costs, or to later structural distress with consequent correction or increased maintenance costs. The geotechnical engineer's construction review services on a drilled pier operation are confined to the drilling, the placement of reinforcing, and the concreting of the drilled pier foundations. He provides the inspection services which comprise an important part of his duties; makes field decisions when the necessity arises; keeps records and makes regular, periodic reports; and provides technical liaison between owner, designer, and contractor.

Note that the geotechnical engineer does not normally perform surveying functions. He does not lay out or independently check the pier locations; and he does not determine or check their as-built positions, unless he has been provided with adequate reference points to facilitate measurements and has been assigned these responsibilities. Proper survey control must be established before the geotechnical engineer can proceed with such work, if assigned.

When drilled pier construction is complete, the geotechnical engineer submits a report stating his opinion of the conformance of the construction to the plans and specifications.

Qualifications   The geotechnical engineer should be a civil engineer or engineering geologist, trained and experienced in the art of soil and foundation engineering. Training in rock mechanics is needed on some projects where loads are heavy and rock conditions present problems, such as pier holes in cavernous limestone or dolomite. Training and experience in engineering geology are very useful in this field, and of great importance on some jobs. Primary training in engineering geology, with supplemental training and experience in foundation engineering, often provides excellent qualifications for this position. (To avoid conflict with the engineering registration laws, a geologist in this position should not be designated as an engineer unless he is licensed as an engineer.) An engineer and geologist, working as a team, often provide very good coverage for geotechnical problems.

Foundation Inspector

Obligations   The foundation inspector is expected to perform the following duties: (1) to be familiar with the plans and specifications; (2) to identify predesignated bearing strata when they are encountered in the hole; (3) to make measurements of hole depth, plumbness, and underream diameter if appropriate; (4) to inspect the hole for cleanness and freedom from excess water before concrete is placed; (5) to observe concrete and steel placement; (6) to note and immediately advise the geotechnical engineer and the owner's representative of any practices used that might adversely affect the compliance of the finished pier with the intent of the plans and specifications; (7) to check the locations of the finished piers against designated locations (but only if directed to do so and if adequate reference points have been provided); (8) to keep daily and weekly records of all pier construction work, including all reports issued regarding unsatisfactory work or construction practices; and (9) to record remedial measures taken and changes made in construction practice. The foundation inspector may be required to sign a report of pay items completed by the contractor, at the end of each work day.

All significant communications to the geotechnical engineer should be in writing, with copies to all concerned.

McKaig [5.4] states:

> The inspector should always remember that the work may become the subject of litigation even though all parties try to avoid such a development. He should therefore maintain his records in such a manner that his testimony, if needed, will be amply documented.

Qualifications The foundation inspector should be able to perform competently the functions listed above to the extent that they are required on the job. On jobs where there is no geotechnical engineer, the foundation inspector should have not only the ability to identify the assigned bearing strata, but also sufficient experience in foundation work to enable him to recognize unfavorable or unanticipated soil or rock conditions and improper construction practices.

In addition to the technical qualifications listed above, the foundation inspector and the geotechnical engineer should have the personal characteristics of integrity and intelligence. A most important qualification for either of these positions is the ability to "get along" with the contractor and his personnel while maintaining a position of objectivity.

5-4 PRECONSTRUCTION PREPARATION

5-4.1 Study of Available Data

The foundation inspector should, before going on the job, familiarize himself with available data that may relate to his work. These will include the plans and specifications, the soil and foundation report, if any, as well as pertinent local building code provisions. In the event that the test boring logs are reproduced on a page of the plans (a practice that should be discouraged), the inspector should study the logs and examine the soil and rock samples, particularly those from the assigned bearing stratum and the strata immediately above and below it. It is the owner's obligation to have these samples available for the inspector's use in the field. These samples should be preserved at their natural water content for this purpose. Many soil and rock materials look quite different after they have been allowed to dry. Rock cores should be surface-wet when examined, when appropriate, to make sure their appearance coincides with that of material as it is brought out by the pier-drilling auger.

5-4.2   Determining Scope of Assignment

The scope of the foundation inspector's assignment varies as mentioned earlier. This scope should be defined in the construction contract, for the contractor's information; it should be defined in the agreement between the firm furnishing the inspector and the owner; it should be set forth clearly in the foundation inspector's instructions, which should be in writing; and—very important—these three documents must be in agreement with one another.

In some instances the pier design is based on support by sidewall shear alone, and only a cursory inspection of the bottom of the hole is required. In other cases the foundation inspector will be required to enter the hole and examine the bottom closely, or to observe the drilling of proof-test holes in the bottom. In some cases there will be a concrete inspector on the job to make test cylinders and slump tests; on other jobs the foundation inspection may include these duties (but only if the inspectors have been properly trained). Other variations in the duties of the foundation inspector are possible, and everyone involved should know what he is not expected to do as well as what his assigned duties are. Often this determination requires a preconstruction conference including the owner, the designer, the resident engineer (if any), the geotechnical engineer, and the foundation inspectors.

5-4.3   Site Visit and Meeting with Contractor

Before a drilled pier operation is started, it is often helpful for the geotechnical engineer and the foundation inspector to visit the site; to discuss with the contractor, the owner's representative, and others involved the inspection duties and their possible effect on the contractor's operations; and to make sure that the contractor will have at the site any equipment that he must furnish to enable the foundation inspector to perform his assigned tasks. As discussed in Chap. 7, the authors prefer to see specifications and contracts so written that a contractor has flexibility in equipment and methods, and is not handicapped by restrictions on either. But there are instances where specific local experience has shown that it is to the owner's advantage to specify a minimum acceptable torque for the drilling machine, for example. The foundation inspector should be aware of any such requirements or limitations on equipment in the plans and specifications. He should report to the geotechnical engineer and have the owner's representative advise the contractor regarding the acceptability of any equipment that has been proposed for use on the job. This should be done *before the contractor brings the equipment on the site.* If it is left until later, rejection of equipment can result in additional expense to the contractor or to the owner.

Another particular point the foundation inspector should check is that the contractor's drilling tools conform to any requirements that are set forth in the specifications.

At the inspector's preconstruction meeting with the contractor, a list of persons who will receive copies of the inspector's reports should be made up, and copies of the list should be distributed promptly to all persons concerned.

5-4.4 Coordination with Structural Engineer and Building Inspector

The foundation inspector (and the geotechnical engineer if the two positions are separate) should meet with the designer of the structure and with a representative of the local building department before the work is started, and the inspector should establish lines of communication (through proper channels) so that he can get prompt action and decisions if questions arise that he cannot resolve.

5-4.5 Written Instructions to Foundation Inspector

Detailed written instructions, with a checklist of equipment to be taken to the job, should be furnished to the foundation inspector by the geotechnical engineer. If there is no geotechnical engineer assigned to the job, this is the responsibility of the resident engineer or the architect. It is important that these instructions should not direct or authorize the inspector to request the contractor to do more than is required or implied by the specifications and contract. The existence of clearly expressed written instructions, discussed and agreed to before inspection starts, is of prime importance not only in performing the job, but also in avoiding or resolving disputes or claims that might arise later regarding job performance.

5-4.6 Inspection Equipment

The foundation inspector may need any or all of the following equipment, some of which will be needed for performance of his inspection duties and some for his comfort and safety:

Folding rule or measuring tape
Powerful flashlight
Hand mirror for reflecting sunlight down the hole
Sampling trowel, knife, jars or sacks, labels
Pocket penetrometer and/or torvane (or other instrument for measuring cohesive strength of soil)

Rock probe (Fig. 5-5)

Geologist's hammer for sampling and testing soundness of rock

Recording and report forms

Safety helmet

Raincoat or rain suit

Safety line and harness

Miner's safety lamp or oxygen test meter

Air tank, fittings, and mask

Of course not all of these will be needed on every job, but they should all be available on any job site where they might be needed, and the foundation inspector should conscientiously check that all needed items are present and in good working order before he begins his work. Some of this equipment—particularly that concerned with safety—is usually provided by the contractor. Any such requirement should be stated in the specifications.

5-5 INSPECTION OF THE HOLE

5-5.1 Safety Considerations—Entering the Hole

The first hazard encountered by the pier hole inspector is that of stones flung from the auger. With the open-helix type of drilling tool, soil and rock cuttings brought out of the hole are spun off the auger, to be picked up and removed by other equipment during the auger's next trip into the hole. Stones of considerable size are often brought out and flung outside the radius of deposit of lighter particles (see Fig. 5-2).

Surface inspection—either sampling and examination of spoil or looking into the hole—should be performed only when the machine operator has been made aware of the inspector's needs and intentions.

Entering the pier hole for inspection of cleanout can be very risky. It is imperative that the foundation inspector take all precautions that might be needed for his own safety, and he should advise the contractor and his employer promptly of any hazards or of any corrective measures that he considers necessary for his safety. *He must never become careless, nor enter a hole when he has the slightest doubt of safety or when any desired precaution has been omitted or skimped.*

Some of the hazards that may be present, and the safety precautions that should be taken, are discussed in the following paragraphs.

Caving or Collapse of Hole or Bell This is the most obvious hazard, and one of the most dangerous. There are some geologic formations

Fig. 5-2 Model SS 8487 Crane attachment, spinning spoil off a large auger. The men stand well back from the hole to avoid being hit by flung stones. (*W-J Sales Engineering Co., Inc., Houston, Tex.*)

of such strength and uniformity that holes and underreams can be counted on to stand open without caving or sloughing. However, if there is no local experience to indicate that this is the case, the inspector should insist on the use of protective casing before he enters the hole. Even where local experience has led to customary inspection and clean-out without protective casing, anomalies are possible, and the inspector should be alert for danger and should insist on the use of protective casing in the shaft if he has any doubts at all or if the holes are of unusual depth or diameter. Slickensided or fissured, overconsolidated clays or slaking shales, which drill cleanly and stand nicely for under-reaming, are subject to dangerous movements of large blocks when exposed to wetting or drying, and particularly when they carry even very small amounts of water in pervious seams or layers, or in joints or fissures.

The protective casing should be positively supported at the top and should extend to within 3 or 4 ft of the bottom, or to the top of the bell, or to such other depth as the inspector may direct.

Fall-in from Top of Hole Before descending, the inspector should make sure there are no loose stones, clods, tools, or even piles of loose soil or rock that could fall or be knocked into the hole while he is in it; and he *must* wear a safety hat. If water is "raining" into the shaft or bell from water-bearing strata, he should wear suitable rain gear. In cold weather this is a safety measure as well as one intended for his comfort.

Gas or "Bad Air" in Hole The authors know of at least one fatal accident from this cause, and have observed many incidents which could have led to fatalities. There are many known possible sources of gas or causes of "bad air." Some of these that the authors have encountered are listed below:

1. Marsh gas from natural organic deposits
2. Gas from organic material in fills (especially garbage fills)
3. Gas from volcanic sources
4. Gas from natural gas or petroleum deposits
5. Gas or fumes from industrial wastes or from leakage of volatile liquids from tanks, pipelines, etc. (This source is not at all uncommon near gasoline stations.)
6. Accumulated CO_2 (sometimes mixed with CO) from motor traffic, or from any internal combustion motors, such as a battery of air compressors. A common source of CO and CO_2 in the hole is the use of a gasoline-driven bottom-hole pump. (These gases, being heavier than air, may also flow into a pier hole from the surface, displacing the air.)
7. Oxygen depletion—for example, from a previous descent, or from too long a stay in the hole [5.5]

Fig. 5-3 Wrist safety harness, designed to allow freedom of movement but to permit lifting an unconscious man without his hanging up on casing. (*Mine Safety Appliances Company, Pittsburgh, Pa.*)

Any of these conditions can be present without giving any warning by odor. Any of them will cause a man to lose consciousness, often without his realizing that he is doing so; and any of them will cause death in a few minutes unless the victim is removed to good air and, in some cases, is given artificial respiration. In every case where the possibility of bad air is suspected fresh air should be introduced into the bottom of the hole by an air hose; and rescue means should be on hand for every descent into a pier hole, *whether it is believed to be dangerous or not.*

A hole can be tested for bad air by lowering a miner's safety lamp or an oxygen test meter.

Safety measures that should be taken for every descent include the following:

Safety harness and line (*Fig.* 5-3) *should be worn on every descent,* even in shallow holes, so that an unconscious man can be hoisted out of the hole without delay and without the danger of injury inherent in the use of improvised rope slings. The practice of sending a man down to lift and carry out an unconscious man, or to hold him while the two of them are being hoisted out, is dangerous and unsatisfactory.

Fig. 5-4 Air mask, reservoir, and harness. (*Mine Safety Appliances Company, Pittsburgh, Pa.*)

Smoking in the hole is unsafe and should never be done.

Air tank and mask should be part of the foundation inspector's kit and should be strapped on and arranged for quick application (not hand-carried) on every descent (see Fig. 5-4).

An observer at the top of the hole should be present at all times when a person is in the hole. It is not sufficient to assign this duty to the drilling machine operator; he cannot properly operate his hoist during a rescue operation and at the same time be sure that the victim is lifted freely and without injury. There should be reliable means of communication between the man in the hole and the observer. In very deep holes a wired or wireless telephone system should be used.

Lighting—An electric light of the safety lamp type, with the cord in good condition, should be suspended in the hole in such a position that the inspector can move it to inspect any part of the hole.

Safety during descent—For descent into holes that are 3 ft or more in diameter and not more than 20 ft deep, a ladder may be the most convenient means; but in most instances some sort of hoisting equipment should be on hand for rescue purposes in case of trouble. It is almost impossible to carry an unconscious man up a ladder in a small-diameter hole.

It is fairly common practice for the drill operator to let a man stand on a rod or pin through a hole in the bottom of the kelly bar and hold onto the bar while it is lowered into the hole, or to sit or stand in a sling or bosun's chair raised and lowered by the drum line on the rig. Unless there is a "fail-safe" arrangement on the kelly hoist or drum line, either of these is a risky procedure. Cable- or chain-operated kellys should never be ridden by personnel, nor should cables on mechanical winches. A positive forward and reverse hydraulic winch is relatively safe; so is a power-up and power-down hoist line on a crane. A hand-operated windlass with a ratchet to prevent accidental release, on a tripod set up over the hole, is usually safe.

The Contractor's Responsibility Whatever means is used for entering and leaving the hole, the contract specifications should state that it is the contractor's responsibility to furnish it; and the conveyance should be convenient, safe, and not uncomfortable for the user. It should also be the contractor's responsibility (and this should also be stated in the contract) to take all applicable safety precautions, including any requested by the foundation inspector, and to furnish any equipment or supplies needed for such precautions (beyond those in the inspector's kit). The need for such equipment is one of the items the inspector should discuss with the contractor before the job is started. Anything that might be required should be on the job site and readily available when needed, and no one should have to hunt for it.

5-5.2   Identification of Bearing Stratum

The plans should indicate the stratum or strata of soil or rock in which it is expected that the pier holes will bottom. Usually, the descriptions will be taken from the soil and foundation report and will be worded so that identification of the various strata as they are penetrated by the auger should be easy. Occasionally, plans will show boring logs made out by someone without geologic or soil identification training, and soil and rock descriptions may be uncertain, or confusing, or even incorrect and misleading. (A common example is the designation of an inorganic silt as a "clay"—a dangerous error because silt and clay behave quite differently under load and during excavation.)

As mentioned earlier, the foundation inspector should examine the boring samples in advance, become familiar with the appearance and "feel" of the proposed bearing stratum, and take typical samples, *sealed in jars at natural moisture content,* to the job site for comparison.

Occasionally, when pier holes are very shallow and when the bottom is clean and there is no water in the hole, it is sufficient to inspect a hole from the surface, examining the sides and bottom by reflected sunlight from a hand mirror or, on dark days, with a powerful flashlight. The condition of the proposed bearing surface can be often judged by probing the bottom with a rod. Identification of the formations penetrated and of the bearing stratum can usually be made from the cuttings as the drilling proceeds. Examination of the larger intact lumps of soil or rock brought out by the auger will give some information about the *in-situ* firmness or strength of the formations. Inspection from the ground surface is not generally recommended, but is the only means available in holes smaller than 24 in. in diameter.

In the normal pier-inspection process, the foundation inspector will have to descend into the holes and examine the proposed bearing surface.

The foundation inspector must always be on the alert for variations in stratigraphy or for soil or rock conditions that are not in agreement with the reported boring information. Most test borings are not sampled continuously, and continuity of strata between borings is never certain. It is not unusual for significant variations in subsurface conditions to be missed in the test borings, but show up in the pier holes and require on-the-spot decisions by the foundation inspector.

The function of identifying the bearing stratum is one of the most important and critical tasks the foundation inspector has to perform. Obviously, the safety of the structure depends on the foundation's resting on suitable support. More than safety, however, depends on the foundation inspector's performance of this task. Drilled pier jobs are priced on the basis of an estimated average depth and a range of depths for the pier holes. Every foot of depth past those shown on the plans will increase the cost of the foundations, often beyond the owner's budget. The foundation inspector must not hesitate to advise the contractor if the right material (or condition) has not been reached, but he must also be positive in identifying suitable support when it has been reached, and he must notify the owner *immediately* if the contractor does not stop at the designated level. Many present-day pier-drilling machines are very powerful, and will cut very competent formations with such ease that there is danger of drilling on past the designated bearing stratum before it is recognized, unless the anticipated change is from soil to hard rock. Some machines will even drill relatively hard rock with comparative ease.

5-5.3   Inspection of Water-filled Holes

It is always preferable to inspect, and to place concrete, in a dry hole; but occasional instances will be encountered where dewatering the hole proves impractical.

For piers in soil, this contingency can be handled by drilling in a mud-filled hole (Chap. 4); but here inspection is limited to observation of drilling, examination of cuttings (which will be mixed with driller's mud), and probing of bottom and sides with a rod. This is incomplete inspection; cuttings will be softened and may be so altered or contaminated as to reduce stratum identification to guesswork. Preparation of pier holes in this fashion is an unusual procedure, to be undertaken only when normal methods cannot be used. The risks from inspection under these conditions can be minimized by care. They cannot be completely eliminated.

In pier holes in rock where dewatering proves impracticable and it is planned to use pumped-in or tremied concrete, inspection can be much more thorough. Rock cuttings can be examined and classified easily during the drilling, the completed hole can be flushed out with clear water, and sidewalls and bottom can be examined in detail by means of a closed-circuit television camera. This means of examination is especially applicable to inspection of holes of great depth.

5-5.4   Bottom Cleanout or Stabilization

Any drilled pier hole should be inspected to assure that there will be good bearing contact between the concrete and the proposed load-bearing surface. Most holes require hand cleaning of the bottom, and holes for piers that depend on sidewall shear for part of their support should be inspected for clay smear on the walls, or any other condition that could interfere with the development of that support. Where grooves in the wall are required, the foundation inspection must make sure that they are properly cut and are clean. Inspection of bottom should immediately precede concreting, because sloughing of bell or shaft wall, fall-in from the surface, or water-softening of an initially sound bottom surface could possibly alter the bearing condition enough to lead to damaging postconstruction settlement.

Ideally, the bottom of a pier hole should be clean. The geotechnical engineer must be the judge of how much loose material—or mud or "slop"—can be left on the bottom without risking later settlement of the loaded pier as the soft material in the contact zone consolidates under load. The foundation inspector's written instructions should give him direction on this point, and he should not hesitate to consult the geotechnical engineer or supervising engineer if he is in doubt about the acceptability of the bearing surface. (See also discussion on pp. 145–146.

When ground conditions are such that the contractor finds it impossible or very difficult to obtain and keep a clean bottom until the concrete is placed, the inspector should immediately consult the geotechnical engineer about a change in procedure or design.

The foundation inspector should report any tendency to leave a hole open longer than necessary, especially in underreamed holes. An experienced contractor is not likely to do this, because it means trouble in cleanout, and under some circumstances can mean complete collapse of the hole or bell, and redrilling to a deeper bearing level at the contractor's expense. Incidents of this nature should be described in the foundation inspector's daily reports, and be reported promptly.

5-5.5   Proof-testing

Rock Bearing The need for proof-testing of rock-bearing surface depends upon the possibility of less competent zones of rock occurring within the depth of effective stress increase below the readily inspectable bearing surface. The need for proof-testing should be determined in advance, if possible, based on knowledge of local subsurface conditions and information obtained in the foundation investigation and presented in the soil and foundation report. If proof-testing is considered necessary, the method and frequency of testing should be detailed in the contract documents, and there should be a provision in the contract for increase or decrease in the estimated work involved. In the authors' practice, proof-testing is usually required for all pier holes in rock formations subject to significant and irregular weathering, where the unit loads to be imposed on the bearing surface are 10 tsf or more. Some limestone formations are so weathered, fissured, and cavernous that every pier hole should be proof-tested, regardless of the design loading.

Proof-testing of rock, as of 1971, usually consists of drilling an exploratory hole, about 2 in. in diameter, in the bottom of a pier hole that

Fig. 5-5 Rock probe rod, made from ½-in. reinforcing bar.

Chisel point

has reached a supposed bearing level, and observing such indicators as speed of drilling under a given drill pressure, dropping or clogging of bit, loss of drilling water (if used), and continuity of the bearing rock as judged by probing (or scraping) the sides of the completed probe hole with a right-angled chisel point formed at the end of a ⅜- or ½-in. steel rod (see Fig. 5-5). Sometimes the proof-test drilling is done with a rotary diamond-coring machine, but more commonly a percussion rock drill is used. With either machine the foundation inspector should see that a sharp clean bit is used, so that the speed of penetration of the drill can be compared between test locations, and that successively smaller bits are used if necessary to prevent drag or clogging as the test hole is deepened.

This method of proof-testing is crude, but fairly effective. In such formations as limestones and dolomites, and the weathered mica schists and gneisses of the eastern Piedmont region, the authors consider it adequate, and necessary for heavy loads or important structures. In the shales of the midwest, this test is not used. In these rocks, caverns and solution channels are very rare or absent, and the rate of penetration of either type of drill is not a good index of hardness because it is governed more by the stickiness of the clayey cuttings than by the hardness of the formation. In these formations, identification of the formation and the depth of "refusal" to the standard penetration test are considered adequate in lieu of any deeper proof-testing.

The foundation inspector should record the type of drilling machine used and the type and condition of the bit. He should explain the desired proof-testing procedure to the driller, record any relevant observations that the driller can pass on to him, and record the time for each 6 in. of penetration. He should decide, on the basis of the criteria established by the geotechnical engineer, when to stop the drilling and make his sidewall probing test. If the probing shows lack of continuity of competent rock, the foundation inspector will state that satisfactory bearing has not been reached and recommend deepening the pier hole.

Instruments and test methods that promise a more scientific approach to rock proof-testing are under development; and it is anticipated that within the next few years more reliable proof-test data will permit design criteria that come closer to developing the real load-carrying capacity of rock foundations.

Cost of Proof-testing   The present (1971) cost of proof-testing in rock by drilling techniques has been found to be of the order of $7 to $15 per test. Because the unit loads that could be approved if the proof-testing were not used would be only a relatively small fraction of the design loads that can be confirmed by the tests, it is obvious that proof-testing (where needed and applicable) is not an extra cost but an economy measure.

Penetrometer or torvane tests in clayey soils should not be considered an extra operation at all. The tests do not require anything of the contractor or delay his operations, and they do not take any more of an inspector's time than the "thumb" test he would make if he did not have the field instruments.

Soil Bearing   An experienced foundation inspector who has familiarized himself with the appearance and "feel" of the test boring samples from the proposed bearing stratum should be able to detect a weaker phase of this stratum from his examination of the auger cuttings and of the bottom of the hole.   However, the use of a pocket penetrometer, a "torvane," or some other field instrument for determining *in-situ* strength of clayey soils is recommended.   The consistent use of such a tool helps keep the inspector's judgment of clay strength from "drifting" and provides a valuable record to back up the inspector's notice to the contractor that bearing level has (or has not) been reached.   A word of caution must be introduced here, however.   Instruments for measuring *in-situ* shearing strength of clays will not give proper indications of shearing strength of cohesionless soils (silts or sands).   An inexperienced or incompetent inspector who uses an instrument of this sort on an inorganic silt stratum, for example, and on the basis of the results approves or advises the deepening of the pier hole, can cause a major (and possibly unnecessary) increase in the cost of a foundation.   Inspection in formations subject to this sort of anomaly requires performance or careful supervision by a geotechnical engineer who can correctly evaluate the effect of the cohesionless layer on foundation performance.

5-5.6   Plate Load and Full-scale Load Tests

On the occasional job where plate load tests or full-scale load tests are required to confirm design criteria, or to justify design changes involving changed criteria, the tests should be designed by the geotechnical engineer, but generally the actual performance of the tests is the responsibility of the contractor.   The number and method of tests should be specified in the contract documents or, if necessary, be covered by a supplementary agreement.   The geotechnical engineer and the foundation inspector should serve as observers on these tests, recording the test data, deciding when load increments should be added or removed, and plotting and analyzing the test data.

5-5.7   Determination of Location, Plumbness, and Dimensions

Alignment and plumbness of the pier holes are the responsibility of the contractor.   To check the agreement of these measurements, as built, with the plans and specifications may be one of the assigned duties of the geotechnical engineer, and these duties are often delegated to the foundation inspector.   Plumbness is easily checked, using a plumb bob suspended from the top of the hole, but alignment—i.e., position of the center of the shaft, at the top, with respect to the design position—often is impossible for a foundation inspector to determine accurately. The geotechnical engineer should not accept this responsibility unless the owner agrees to furnish batter boards (or other suitable location markers) that are not disturbed by the pier-drilling and concreting operations. Such markers should be so positioned as to allow the needed measurements to be made on the seated casing and on the completed piers. It is the contractor's responsibility to see that the batter boards or other location markers are not disturbed.

When an uncased pier has been concreted, it may be impossible for the inspector to determine the exact location of its center, for such pier holes are rarely completed without some irregular raveling at the surface. The remedy is to set a short piece of surface casing or liner, to be left permanently in place, before the pier is concreted.

Insofar as is practicable, the foundation inspector should inform all parties as soon as it appears that a drilled pier hole does not conform to the specified tolerances, so that corrections can be made before concrete is placed. He should notify the owner's representative (through the geotechnical engineer) immediately whenever it is determined that a completed pier is not correctly located, so that design of corrective piers or other measures can be started. Any avoidable delay in getting such corrections started might be used by the contractor as the basis for a claim for an "extra."

Occasionally ground conditions will be such that the contractor will be consistently unable to complete piers within plumbness or location tolerances. As soon as this condition is noted, it should be reviewed with the geotechnical engineer and the owner's representative, so that the possibility of relaxing the tolerances, or of redesigning the piers to permit larger tolerances, can be considered. A few inches increase in shaft diameter, for example, might be less expensive and more acceptable to all concerned than the use of additional piers to compensate for misalignments.

5-5.8 Guarding Against Lost Ground

A drilled pier hole in ideal ground conditions does not produce appreciable movement of adjoining ground. There is a tendency to think of all drilled pier installations as being in ideal ground conditions, probably because contractors are increasingly able to cope with unfavorable ground conditions. The potential for lost ground is often overlooked until subsidence has occurred, sometimes with serious damage to nearby buildings, streets, or utilities.

The foundation inspector should be alert for ground conditions that could produce inflow of excess soil as the drilling is done, and for detrimental effects of groundwater drawdown. He may also be instructed to examine nearby surfaces and structures before pier drilling is started.

If he has this duty, existing cracks in buildings, ground, or pavement should be documented by dated and signed photographs, sketches, and notes, which should be part of the foundation inspector's first reports. For major works, this documentation requires professional photography and surveying.

Inspection for evidence of lost ground should be repeated daily. Indications of lost ground may take the form of subsidence immediately adjacent to a pier location, new or widening cracks, curb separating from pavement, soil pulling away from nearby foundation walls, or development of sags or "bird baths" in adjacent pavement or lawn. An excess volume of soil may not be evident as it is removed from the hole; but there may be indications during the drilling noticeable to the drilling-machine operator, such as closing in of the hole, sand inflow, or soil in the water discharged from dewatering pumps, whether they are pumping from a pier hole being dewatered for concreting, from a deep-well drawdown system, or from a well-point installation.

The foundation inspector should also observe any groundwater variations that occur during construction and be alert to the possibility of ground movement due to such variations.

At the first indication of potentially damaging ground loss, the foundation inspector should make measurements and notify the geotechnical engineer, so that recommendations can be made for changes in construction procedures or other special measures can be taken to prevent damage to existing structures.

5-6   REINFORCING BAR CAGE INSPECTION

Inspection of reinforcing bar cages for size and condition of bars and dimensions of cage is not usually included in the duties of the foundation inspector. However, the correct *positioning* of the cage in the pier is an important matter, and the foundation inspector is on hand when the cage is placed in position and while the concrete is being placed. The contract documents should state clearly who is responsible for observing and reporting on the positioning of the cage, the means by which it is held in position, and its apparent position after the concrete has been placed.

5-7   INSPECTION OF CONCRETE PLACEMENT

5-7.1   General Objectives

The general objectives of concrete inspection on a drilled pier job are (1) to see that the concrete has the specified strength; and (2) to make sure that the concrete of the pier is continuous, in full design section, from the bottom to the top of the pier. The first of these objectives is common to all concrete jobs. The task of taking test cylinders and making slump tests may be assigned to a foundation inspector, but is more commonly contracted to a commercial testing laboratory that has the responsibility for quality assurance of the concrete. The requirement of assuring integrity and continuity of the finished pier is extremely important. *It requires specialized experience and knowledge and should be handled by someone qualified as a foundation inspector.* When a drilled pier is completed, whatever has been done is buried. There is no opportunity later to strip forms and see how it looks or to use a "soundness hammer" on the sides of the column to check the integrity of interior concrete. Although observation and reporting of the placement of concrete may be an assigned duty of the concrete inspector, *this function is so important that it should also be a responsibility of the foundation inspector.*

5-7.2   Water in the Hole

If possible, the pier hole should be clean and dry when concrete is placed. In many instances, however, complete absence of water in the hole will be impossible. This contingency should be anticipated in the design and covered in the specifications. The allowable depth of water will depend on (1) the type of pier, and (2) the method of concrete placement. Two in. of water in the bottom of a straight pier hole can usually be displaced or mixed with free-falling concrete without causing serious segregation or weakening of the concrete; but the same amount in a belled hole will have much greater effect, and may, as it is displaced and rises into the shaft, cause complete loss of strength in the concrete at the junction of bell and shaft. The use of clean coarse gravel, or of dry cement, to "blot up" a limited amount of free water is sometimes permitted, but either device should be used with caution.

When concrete is placed by tremie, by "elephant's trunk," or by pumping, with the hose at the bottom of the hole, free water in the hole has little effect.

The foundation inspector must know what water depths the specifications will allow, as well as the method of placement specified. He must be prepared to insist on (1) effective dewatering measures, or (2) tremie placement or pumped-in concrete, in case excessive water enters the hole. Approval must be obtained from the geotechnical engineer if deviation from the specifications is involved, and the approved deviations must be recorded in the foundation inspector's written reports.

Before concrete can be placed under water, either by tremie or by pumping, the water level in the hole must have reached a stable level; otherwise, water will continue to flow into the hole, washing or diluting the concrete, until the concrete level has reached a point where its pressure exceeds the outside water pressure at the point of entry.

The specifications should specify a minimum size of tremie pipe, and the inspector should see that this limit is observed. If this item is not covered in the specifications, he should not approve the use of a tremie pipe smaller than 8 in. in diameter, and 12 in. is preferable. (Note that this limitation applies to the pipe for *tremied* concrete, but not to that for *pumped* concrete.) The inspector should keep in mind that tremied concrete will add $12 to $14 per cu yd to the cost of the pier, and that a good casing seal and dry hole are a better solution where they can be obtained economically.

5-7.3   Final Inspection Before Concrete Placement

Although a pier hole may have been inspected and found clean and dry and ready to be filled with concrete, if more than a few minutes have passed after inspection, it should be inspected again immediately before the concrete is placed. Unexpected sloughing or caving of a bell or uncased shaft can occur. There have been serious instances of deep collapse of permanent casing produced by soil and water pressure, undetected until the pier had been completed and partly loaded [4.29]. Immediately before concrete is placed, the foundation inspector should take a final look at walls, bottom, and reinforcing cage, using a powerful flashlight.

5-7.4   Cold-weather Concreting—Fog in the Hole

Often when a deep drilled pier is to be completed in cold weather, fog will form in the hole and interfere with the final inspection from the top. This may occur before concrete placement is started. It is likely to happen as soon as the relatively warm concrete begins to enter the hole. In either case, the presence of fog must not be accepted as an excuse for omitting or skimping on the final inspection of the hole and the observation of concrete level as the hole is filled or as casing is pulled. If necessary to allow inspection, the fog must be dispersed by lowering a heater or blowing warm air down the hole. If cold-weather concreting of deep piers is anticipated, this item should be provided for in the specifications; otherwise the contractor may be unprepared and be entitled to an "extra."

5-7.5   Avoidance of Segregation

To avoid detrimental segregation in concrete, the specifications usually provide that the concrete shall be discharged through a hopper having a bottom spout centered over the hole to concentrate the falling concrete in a stream of small diameter compared to the hole or reinforcing cage diameter, and that the hopper shall never be permitted to empty until the hole is filled with concrete. The foundation inspector must see that this provision is followed, or that a satisfactory alternate method of avoiding segregation, such as the use of an "elephant's trunk," a tremie pipe, or pumped-in concrete, is employed. If an alternate method is used, it must be approved in writing by the owner's representative (or the geotechnical engineer) and must be documented in the foundation inspector's reports.

5-7.6   Concreting in Reinforcing Cages

Concrete slump and maximum size of aggregate must be carefully designed to assure that either free-falling or pumped-in concrete will flow freely between the reinforcing bars and completely fill the space outside the cage. Although these are details normally covered by the structural engineer and his concrete inspection agency (usually a commercial testing laboratory), they are so vital to the performance of the pier that the foundation inspector too should be alert to report the placement of any concrete that does not flow freely through the openings in the cage. The foundation inspector will judge this by the appearance and behavior of the concrete and by the volume of concrete required to fill the hole. Concrete with less than 4-in. slump will not flow readily between closely spaced reinforcing bars, and a properly designed mix with a 6- to 7-in. slump is preferable. The inspectors should never accept concrete that has had water added to it to increase its workability.

Under some circumstances, reinforcing cages can become displaced or distorted during concrete placement or during the pulling of temporary casing (see Art. 4-5). This fault may appear as a rise of the cage above its design position, or the cage may "squat" and even disappear below the surface of the concrete (Fig. 4-29).

A foundation inspector must be alert for the conditions that lead to this type of problem, and should if possible watch the top of the cage during removal of the casing as well as watch the concrete surface in the casing. When long lengths of casing are used, direct observation becomes impossible as the casing is hoisted; and the foundation inspector will have to form his judgment from his observation of the top of cage and concrete before and after removal of each casing length, the smoothness of the casing pulling operation, and the fluidity of the concrete. If a reinforcing cage disappears or is obviously distorted or displaced, the geotechnical engineer and the owner must be notified at once and instructions for correction be transmitted from the owner to the contractor.

5-7.7 Vibration

The extent to which vibration is to be used may be covered by the specifications. If it is specified or used, it becomes one of the items that both the concrete inspector and the foundation inspector must observe and report.

Vibrators generally are either electric or air-operated; and models of both are available for use in deep holes. The foundation inspector should include in his report a statement of the kind and model used, its operating condition, and how it was used—at what depths, length of time in operation, etc. It is not sufficient merely to note "vibrated" unless a prior report gives the pertinent details.

5-7.8 Interrupted Pours

Good practice requires, and the specifications should require, that a pier hole be filled in one continuous pour. The specifications should provide for a special procedure to be followed when this cannot be done. Occasionally, due to causes beyond the contractor's control, concrete placement will be interrupted. Procedures for assuring continuity of the pier concrete in such cases are described in Art. 4-5, and the foundation inspector should be familiar with these procedures, in addition to the specific wording of the specifications. The foundation inspector should observe and record these operations and the time at which they occur, and he should refuse to approve any pier in which they are not performed as specified.

5-7.9 Pulling Temporary Casing

This is an operation that is most critical, and one in which probably more mistakes are made than in any other in drilled pier construction. The foundation inspector must be present, alert, and aware of the potential for trouble inherent in the operation and in the particular formations that have been cased off.

When holes have been cased through formations that normally stand open, temporary or protective casing may be pulled after the bottom has been cleaned and inspected but before the concrete is placed. In this event, a clean hole should be exposed and the concrete should be placed immediately, and the inspector has only to watch for sloughing before or during the concrete placement. But in most instances, casing will not be pulled until it has been partially filled with concrete; and to assure the continuity of the completed column of concrete, *the inspector has to watch for any occurrence that threatens the integrity of the completed column.*

Especial care is needed when there is a groundwater level outside the casing, which must be held back by the fluid pressure of the concrete inside the casing when the bottom seal is broken.   Construction errors that can be made under these circumstances are discussed in Art. 4.5.

The foundation inspector should make every effort to know where groundwater level is outside the casing, and if there was any inflow from a pervious stratum (or rock fissure or cavity) during drilling, he should be aware of it and should report it.

The casing *must* be filled with concrete to a level at least high enough to balance the groundwater level outside the casing before it is lifted off its bottom seal.   (The inspector should keep in mind that 2.4 ft of water will be balanced by 1 ft of concrete.)   This will be sufficient to keep a positive head of concrete against inflow of water at the bottom of the casing as it is being raised.   The relation between required concrete pressure and existing groundwater pressure is illustrated in Fig. 4-29.

A point of caution: The specifications should require that temporary casing—especially telescoping casing—be clean and free of dents, or else there will be a tendency to hang up and bind during removal. Some specifications require also that the casing should be well oiled. (Some contractors insist that oiling is a waste of time, as all oil is wiped from the metal surface by the time the casing is inserted and concrete placed.)

The foundation inspector should call the contractor's attention to any deviations from these specification requirements, and should indicate such notice, with a record of compliance or noncompliance, in his daily reports.

Vibratory pile-pullers have recently been adapted to pulling drilled pier casings and have been found very effective.   With vibration, casing can be lifted smoothly at a controlled rate, and the vibration probably helps develop good sidewall shear contact and good bond between concrete and reinforcing steel.   An added dividend—as far as the contractor is concerned—is that the use of the vibrating puller reduces the danger of overstraining crane booms, and thus increases job safety.

5-8 OBLIGATIONS OF OTHER PERSONS

For a smooth-running drilled pier job, there are obligations and qualifications which appertain to other persons as well as to the geotechnical engineer and the foundation inspector.   As the authors see these, they are as follows.

5-8.1   Owner

The owner has an obligation to employ competent engineers and inspectors, not on the basis of cost but rather on the basis of qualifications. He should define their obligations and duties clearly, and back up their decisions and recommendations promptly.

5-8.2   Contractor

The contractor's prime responsibility is to complete the contract on time, in accordance with the plans and specifications, and in a workmanlike manner. One of his obligations, which should be explicitly set forth in the contract, is to correct promptly all specification deviations reported by the owner's representative. Another is to do his work in such a way as to minimize engineering construction review and inspection costs. Experience with contractors readily separates those who can be expected to turn out a well-managed job from those who can be expected to end the operation in avoidable disputes, arguments, and claims for "extras" or "changed conditions." Some of the qualifications that the geotechnical engineer looks for in a drilled pier contractor are adequate work force, suitable equipment, freedom from overcommitment on other contracts, experience with drilled pier construction under conditions similar to those at hand, willingness to make corrections and to change unsatisfactory construction practices when they are called to his attention, and a record for finishing jobs on time and with a minimum of disputes or claims for "extras." (Of course, disputes and claims for "extras" are not necessarily the contractor's fault. Sometimes the grounds for disputes and claims are written into the contract; sometimes they are incorporated in the plans and specifications; sometimes they show up as unexpected subsurface conditions; and sometimes—unfortunately—they are the result of bad decisions by engineers or inspectors.)

5-8.3   Building Department Inspector

This official has the obligation of seeing that any construction in his jurisdiction meets the requirements of his building code and, often, the requirements of departmental rulings or practices that may comprise interpretations of the building code. His duty is to protect the public's safety. One of his obligations is to make his inspections promptly when notified. In order for him to do this, of course, someone must keep him informed of the progress of the work and give him adequate notice of occasions when the work will be ready for his inspection. For drilled pier construction, the foundation inspector should keep this in mind. A good foundation inspector will earn the confidence of the building inspector.

5-9   CONTROL OF THE JOB

As has been indicated many times in the preceding pages, the construction of drilled pier foundations requires suitable design, careful construction, and good and knowledgeable control of construction details if both the supporting capacity and the potential for economy of this type of foundation are to be realized.   This is true of any foundation construction, of course, but both the supporting capacity and the cost of drilled pier construction are more sensitive to variations in these aspects than are most other types of foundation.   Mistakes in design or in construction details can easily be overlooked until too late, resulting in unnecessary additional costs and construction delays.   On projects involving drilled piers, both design and construction decisions (and changes) may be required as the work proceeds, and therefore close job control is imperative.   This means, of course, that the geotechnical engineer and the foundation inspector must be competent.   It is equally important that the contractor must be cooperative (as well as capable).   Finally, the owner should realize the importance—to him—of what the construction review team is trying to accomplish, and should be prepared to take immediate steps to back up their decisions if the necessity should arise.

To implement close control, the inspector has to report promptly and regularly.   The lines of communication between inspector, geotechnical engineer, structural engineer or architect, and owner have to be established and maintained; and both the geotechnical engineer and the inspector have to earn and keep the confidence of the owner and his agents and the respect of the contractor.

A message to owners: competent engineering construction review and inspection do not increase job costs.   They save both money and time.

5-10   POSTCONSTRUCTION INSPECTION

There have been some large-scale construction errors in recent years which resulted in the necessity for replacing some very large and expensive piers in which the completed concrete shaft was found to be discontinuous or locally reduced in cross section.   As a result, it has become customary in some areas to check the integrity of a representative sampling of completed piers on important jobs by making diamond core borings for the entire length of the shaft; or, in some cases, by making a boring outside but parallel to the shaft in order to confirm the presence or condition of the bell [4.17]. This test is applied also to piers whose integrity is suspect because of construction circumstances or inspectors' observations. (See also Art. 4-6.)

REFERENCES

5.1 Loss Prevention Manual, *Consulting Eng. Council U.S.A.*, 1969.
5.2 Davidson, David McL., The Legal Implications of Quality Control, *Civil Eng.*, November 1967.
5.3 Goldbloom, Joseph, Safety in Construction—Whose Responsibility? *Civil Eng.*, November 1969, p. 42.
5.4 McKaig, Thomas H., "Field Inspection of Building Construction," McGraw-Hill Book Company, New York, 1958.
5.5 Poisonous Subsoil Air, *ENR*, Aug. 26, 1971, p. 13.

CHAPTER SIX

Exploration for Drilled Pier Foundations

6-1 GENERAL REQUIREMENTS

The authors' experience has led to the conclusion that more careful and thorough foundation exploration is required for drilled pier foundations than for most other deep-foundation systems. Part of this conclusion derives from the sensitivity of drilled pier costs to apparently minor geologic variations, and part from the difficulty of anticipating pier construction contingencies from the results of conventional test borings. This conclusion is probably not correct for areas where the bearing formations for drilled piers are uniform and readily identifiable rocks or soils, and where drilling conditions are generally favorable.

It is imperative that the exploration program include enough borings, soundings, or test pits to establish a strong inference regarding the continuity (or lack of continuity!) of the formations penetrated throughout the area in which the pier holes will be drilled. The test borings must extend into the proposed bearing stratum a distance sufficient to establish the adequacy of that stratum within the depth of significant stress increase imposed by the pier.

The person supervising the exploration program needs both knowledge of the proposed structure and understanding of the behavior under load—as well as during drilling—of the soil and rock formations penetrated. For the benefit of the foundation contractor, good representative samples must be obtained and preserved at natural water content, for inspection later by prospective bidders. "Undisturbed" samples may have to be taken for strength and consolidation testing. The borings should be logged accurately and completely, in such terms that the prospective bidder can recognize the materials penetrated and anticipate their behavior under his equipment.

Because of the important influence of apparently minor variations in materials, continuous sampling may be necessary in selected borings. "Undisturbed" as well as frequent drive sampling is appropriate within cohesive soil strata that might be expected to stand open without casing, and especially within strata in which underreams may be considered.

6-2 CONVENTIONAL EXPLORATION

Any type of test boring that is suitable for retrieval of uncontaminated soil and rock samples can be used; but especial care should be taken in the logging, the preservation of samples, and the accurate determination of groundwater levels. Because of the latter requirement, auger borings through the overburden, using hollow-stem flights where unstable soil conditions are anticipated, are especially suitable. Chop-and-wash and rotary borings, using either water or drilling mud to carry out the cuttings, are likely to have a tendency to seal the sidewalls of the test hole and delay correct indication of true groundwater level, or may obscure detection of water-bearing strata.

Soundings can be used to augment the test borings in establishing the depth to rock or hardpan, the thickness of hardpan or of a weathered rock zone over bedrock, and sometimes the condition of the rock. For this purpose, soundings can be unsampled auger, wash, rotary, or air-drilled holes, which may be carried into rock; or, in some circumstances, probes to rock using a hammer-driven or hydraulically pushed rod, often equipped with an expendable cone point larger than the diameter of the rod.

Test pits can be made quickly and cheaply, using a backhoe or a pier-drilling machine. The excavation depth of backhoe equipment is usually about 10 to 12 ft (3 to 3.7 m). Test pits by a pier-drilling machine can go to any depth permitted by soil conditions, and are especially useful as drilled pier prototypes in defining caving or water inflow problems (see Art. 6-4).

6-3 GEOPHYSICAL METHODS FOR LOCATING
TOP OF BEDROCK

Seismic traverses using refraction methods can be used to supplement test hole data in determining the depth to and the configuration of a buried rock surface, provided there is a marked difference in the seismic velocity of the rock and the overlying materials. Under favorable conditions, conventional multichannel seismographs will enable rock surface depth to be interpreted at eight to ten locations per day. The recorded rock velocities will also provide some insight into the quality (and drillability) of the rock. The seismic method, however, is often precluded in urban locations because of existing underground facilities and background vibrations.

The resistivity method of exploration, most often using the four-pole Wenner configuration, has proved to have some utility in definition of discontinuities within the rocks, e.g., solution voids or major fractures. Although this method is one of the most economical exploratory techniques, the authors have found the degree of success to be mixed and to be directly related to the regional experience and "art" of the operator/interpreter. As with the seismic method, borings are required to provide a "calibration" of resistivity profiles and constant depth traverses.

A relatively new method for determining definitely when test borings have reached bedrock (as distinguished from a boulder or an isolated fragment of hard rock in a weathered zone) has been reported recently by Lundstrom and Sternberg [6.1]. The method merits serious consideration.

A boring is drilled, definitely penetrating 10 ft or more into rock as confirmed by careful logging or rate of drilling, drill behavior, etc. (or by rock core). Normally, a percussion-type, air-operated rock drill is used; penetration is rapid, and this type of test boring is inexpensive. A hydrophone (a nondirectional underwater type of microphone) is set in the hole below rock level. Other test borings (or probes) are made at selected locations, and the sound transmitted during drilling is reproduced at the receiver-recorder connected to the hydrophone. It is reported that it is easy to distinguish, either from the sound or from the shape and amplitude of the recorded sound-trace, between the drilling of continuous bedrock (either igneous or sedimentary) and of an isolated rock element, even when the latter is a very large boulder in contact with the surface of bedrock. An outstanding advantage of the method is that the test probes need only reach the top of sound rock, with no necessity for confirming by further penetration.

The radius of exploration for one hydrophone setting is said to be about 100 m (330 ft).

It appears that this method offers a relatively inexpensive, rapid, and accurate method of determining depth to sound bedrock for projects where foundations must rest on or in sound rock, and where overlying "hardpan," boulder till, or weathered rock make determination from penetration resistance alone difficult or expensive.

6-4 LARGE-DIAMETER AUGER BORINGS
FOR EXPLORATION

Because of the importance of being able to predict soil and rock behavior during pier drilling from the exploratory test results, it is often advantageous, and sometimes imperative, to drill prototype pier holes as a supplement to the conventional subsurface exploration. These test holes do not necessarily need to be the full diameter of the proposed pier holes; but they should be large enough that water entry, caving, sloughing, or squeezing in can be observed and logged accurately as to depth and stratum material and thickness. In some cases this can be done with a hole of only 12 or 16 in. in diameter (30 to 40 cm), logged from the surface; in others, particularly in holes of great depth or at sites where piers are to carry heavy loads, a complete and reliable log can be produced only by foot-by-foot inspection of the completed hole. This requires a boring with a diameter of at least 24 in. (60 cm), preferably larger, protective casing (in some cases), and all the safety precautions needed when anyone enters a pier hole.

It should be remembered that a poorly backfilled large-diameter test boring, if it is too close to a final pier location, can cause construction difficulties, or even unsatisfactory pier performance. For this reason it is prudent to limit test borings within the boundaries of the proposed structure to the conventional small-diameter variety, and to locate the large-diameter test borings outside the structure limits—which should be known and accurately located at that time. In any case, large-diameter test borings should be backfilled with care to forestall trouble of one kind or another at some future time. The use of a soil-cement slurry, a very lean concrete, or a mixture of dry sand with a little cement should be considered for backfill in any location where there is any possibility of later interference with pier-drilling operations or with distribution of loads from completed piers.

6-5 SCOPE OF EXPLORATION

As a supplement to test borings, rock soundings using air-drilling methods have been found particularly advantageous at sites where depth to rock is extremely irregular, where there are potential solution cavities, or where the quality of the rock is variable, and where piers are planned as end-bearing on rock or as "rock sockets." A detailed conventional exploration of irregular or problem rock areas need not involve too great an extra expense, because usually most of the test borings can be sampled intermittently and can be supplemented by relatively inexpensive unsampled air-drilled holes. For some jobs, uncertainties can be reduced, and occasionally even substantial economies can be made by extending the exploration so that a conventional test boring or a sounding is made exactly at each column location.

Test borings should be continuously cored in rock (or drive-sampled in soft rock) for a sufficient depth to investigate all rock within the depth of significant stress increase. A record of the rate of advance (feet per minute, etc.) is also helpful in assessing the drillability of the rock. The detailed information on rock depth and hardness thus made available to prospective bidders for pier contracts will help avoid the need for substantial "contingencies" in the bid prices. The potential saving may be many times the cost of exploration.

A point requiring special attention in test borings where drilled pier foundations are a possibility is the possible need of sealing test holes after they are completed. Occasionally, where a pier location has coincided with the location of an earlier test boring, there will be water entry from a lower aquifer through the abandoned test hole, and the contractor will be unable to dewater the pier hole. This is a serious matter; at the very least, it will produce a delay in construction, and on several occasions it has led to the abandonment of a very large drilled pier and the substitution of another type of foundation [6.2, 6.3]. When it is known that drilled piers may be considered, exploratory borings should be plugged from the bottom up to a level somewhat above possible pier bottom. Satisfactory plugs can be formed by dropping in, and tamping, bentonite pellets (sold commercially as Pi-pellets—see Appendix C), as is done for some types of piezometer; or the hole can be sealed by grouting from the bottom up. Grout for this purpose can be neat cement, cement-bentonite mix, or one of the chemical grouts. A grout that seals but does not offer much resistance to the auger is best.

For sites where depth to rock (or other bearing strata) is known or suspected to be variable, and exploration has not been extensive enough to provide a realistic prediction of pier qualities, it is helpful to consider the drilling of the piers themselves as part of the exploration, *allowing for possible variation in depths or rock conditions in the plans and specifications and the bid documents.* In such cases, the sequence of pier drilling should be planned to provide the needed information as the work progresses, rather than starting at one side of the site and proceeding systematically to the other side, as the contractor might prefer to do. In this way, the presence of a major anomaly may be detected, and the contractor can move on to another area (or back to a proved condition) while the designers are making provisions for any adjustments or redesign that might be required for the unexpected condition. To avoid argument, the contract should provide that the geotechnical engineer can designate the sequence in which pier holes are to be drilled. Whenever actual pier holes are used in this way for continued exploration, they should be logged in detail by the foundation inspector or the geotechnical engineer—a procedure not ordinarily applied to pier holes.

6-6 GROUNDWATER OBSERVATIONS

Not only should the level of water entry be noted on the exploration logs, but the circumstances and rate of water entry are of major importance in planning a drilled pier job. Each exploration hole should be used to obtain as complete coverage of this aspect as possible. The boring log should show the thickness and kind of formation yielding the water (e.g., sand and gravel overlying bedrock; weathered, broken, or solutioned rock; thin layers of clean sand within a clay formation; or fissured clay). Rate of water entry should be estimated or measured. On large projects where groundwater problems are anticipated, meaningful measurements can be made conveniently by using cased borings or piezometers with the tip seated at the level of a potential aquifer, by making pump-in (falling head) or pump-out (rising head) tests, or by measuring stable drawdown or pump-in level at a constant rate of pumping. Techniques and data treatment for field permeability testing in both rock and soil formations are summarized by the U.S. Bureau of Reclamation [6.4] and by Hvorslev [6.5]. Seepage quantity estimates for full-scale pier holes made from pumping tests in small-diameter holes are of much more limited reliability than those made from large-diameter holes, but they can be very useful when larger-scale tests are not feasible, and may provide more specific data concerning individual aquifers.

6-7 ROCK CORING

Whenever a rock formation is considered for pier support—either in end-bearing or in sidewall shear—the rock that is proposed for support should be investigated by core boring. Depth of coring will depend on the size of the pier or underream, the magnitude of the proposed load, and the nature of the rock formation. For end-bearing on rock, coring should extend below proposed bearing level to a depth of at least twice the bearing surface diameter. If poor recovery, or examination of recovered core, indicates the presence of voids, clay, or compressible zones within the interval cored, then coring should be extended further until a suitable continuity of sound rock is encountered. If the pier is to be supported by sidewall shear (a design that may be dictated by the condition of the rock penetrated), then depth of coring should be determined by the geotechnical engineer on the basis of conservative assumptions as to permissible shearing loads (see Appendix E).

Good core borings in rock are expensive and require technical skill, experience, and good equipment. The authors believe that a negotiated contract with the best available drilling contractor is a better investment than letting a test boring contract to the lowest bidder. By careful planning of the use of large-diameter auger holes (to allow preestimation of the penetration possible during actual pier drilling), correlated with carefully logged probes, core boring footage can be minimized.

Unless the rock to be cored is relatively sound and free from discontinuities that could influence core recovery, cores for drilled pier exploration should be of NX size (2⅜ in., or 60 mm) or larger, particularly where compression testing is contemplated. As a general rule, the smaller the core diameter, the poorer the recovery. Often an AX or BX core will suggest poorer rock quality than actually exists, and may result in an initial estimate of deeper pier penetration into rock than is really necessary or economical.

6-8 ROCK DRILLABILITY

The correlation of conventional rock property data and small-scale test drilling (microbit tests) with the large-scale drillability of rocks *in situ* has long been the goal of researchers concerned with drilling for mineral and groundwater exploration and development, and is most recently being studied for application to tunneling by mechanical moles. It is unfortunate that almost no information on correlation of the penetration of large-diameter drills with material properties or with drill test parameters is available in the current literature. Thus, the application of such parameters to assess rock drillability for large-diameter drills is virtually unexplored, although some qualitative trends can certainly be established from the available small-bore research.

Perhaps the most definitive work to date involving small-bore drillability has been in the correlation of microbit drill tests with full-scale drilling, particularly with rolling-cutter bits. An example of the use of microbit testing to estimate the drillability of various rock types with a Tricone rock bit is shown by Fig. 6-1. For a comprehensive view of research applicable to small-bore drilling, the interested reader is referred to a report published by the Rock Mechanics and Explosives Research Center of the University of Missouri [6.6].

Most researchers have found that of all the rock properties considered, compressive strength of representative core samples is of the most value in predicting drillability. However, it has also been demonstrated that

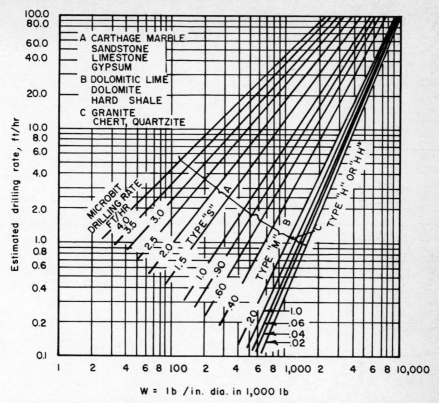

Fig. 6-1 Estimated drilling rate of Hughes Tricone rock bits at 60 rpm as determined by microbit drilling tests.

strength, or any other rock property, is not always a reliable guide. The influence of compressive strength on the drilling rate of various bit types is shown by Fig. 6-2. It is noted that for a given drilling thrust, the drilling rate generally increases with a decrease in compressive strength.

For the various types of drills and drill bits, drillability of rock formations is dependent on many factors which are related to such tool variables as speed of rotation, thrust, and torque, which in turn are related only indirectly to rock properties. It may therefore not be possible to derive simple, representative drillability relationships between drill characteristics and rock properties, particularly for large-diameter borings. However, even rough qualitative correlations between drillability and simple, small-scale test parameters are urgently needed to permit a more rational assessment of drill selection and of the economic feasibility of large-diameter rock drilling. Such assessments would go far to improve the confidence of the engineer and contractor in the use of

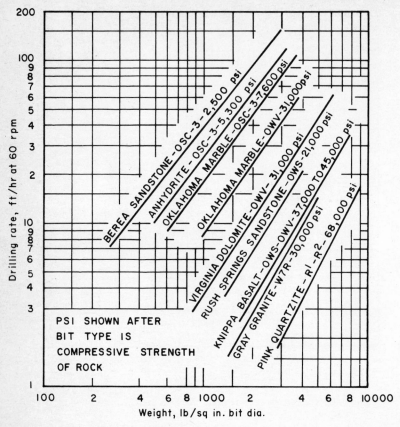

Fig. 6-2 Drilling rate vs. weight on bit for new rock bits. Note that correlation with compressive strength of rock is irregular, but shows a general trend.

piers drilled into hard rock formations to sustain the large loads that are now being imposed with increasing frequency in urban construction.

6-9 RATING OF ROCK QUALITY

For jobs involving very heavy column loads where, for any reason, it is desired to take full advantage of the load-bearing ability of a rock formation, an estimate of rock quality—in other than subjective terms—becomes very desirable. As discussed in Art. 3-3, quantitative procedures have been proposed by Deere [3.21] for rating rock quality by using measurements of fracture frequency (FF) or rock core recovery (RQD), or, alternatively, measurements of field seismic and laboratory sonic velocities. The authors recommend that rock quality indices be determined as part of any exploration program that involves rock support of important structures, a procedure which they follow.

Of the various rock quality indices, the easiest to determine is the rock quality designation (RQD)—the percentage of core in an NX (or larger-diameter) core run which is made up of core segments 4 or more in. in length, measuring only between natural core separations and not between separations due to drilling effects. An alternative, and perhaps the most representative, quality index is the square of the ratio of field seismic velocity to laboratory sonic velocity, as determined from representative core specimens. An example of the utility of rock quality ratings is shown by Fig. 3-9. It is interesting to note that the RQD and the velocity ratio index tend to have a 1 to 1 correlation.

REFERENCES

6.1   Lundstrom, R., and R. Sternberg, Soil-rock Drilling and Rock Locating by Rock Indicator, *Proc. 6th Intern. Conf. SM & FE, Montreal,* 1965.

6.2   Tomlinson, J. M., Discussion of Paper by Palmer and Holland, Session B, Large Bored Piles, *Proc. Symp. Inst. Civil Eng., London,* 1966.

6.3   Toms, A. H., Discussion of Session B, Large Bored Piles, *Proc. Symp. Inst. Civil Eng., London,* 1966. (See also closure of same discussion by D. J. Palmer.)

6.4   U.S. Bureau of Reclamation, "Earth Manual," Designations E-18, E-19, E-36.

6.5   Hvorslev, M. J., Time Lag and Soil Permeability in Groundwater Measurements," *Corps Eng., Waterways Expt. Sta., Vicksburg, Miss., Bull.* 36, 1951.

6.6   "Rock Properties Related to Rapid Excavation," report by Rock Mech. Expl. Res. Center, University of Missouri, Rolla, Mo. (U.S. Dept. Transportation Contract 3-0143).

CHAPTER SEVEN

Drilled Pier Specifications and Contract Documents

7-1 INTRODUCTION—GENERAL CONSIDERATIONS

The specifications and contract for a drilled pier project should give the contractor as much leeway as practicable in the choice of equipment and methods. A performance type of specification, imposing no unnecessary restrictions but requiring an acceptable end result, will in the long run result in more competitive bids, lower bid prices, fewer disputes and claims for extras, and usually better foundations than would a tight specification that unnecessarily restricts the choice of contractors or impedes the work of the successful bidder. For example: Specification of a 30° bell (see Fig. 1-1) might eliminate bidders having only equipment for 45° bells or dome-shaped bells, which are just as good; or unnecessary specification of a minimum torque for the drilling machine could discourage bidders with slightly smaller machines who had learned to drill the specified sizes of holes or bells in the site formations by the use of special techniques or special drilling tools. There will be occasions when it is necessary to limit the contractor in such matters, but the limitations should not be imposed unless the need is real and important.

A specification that is "too tight"—i.e., which imposes unnecessarily difficult or restrictive conditions on the contractor—may lead to bids which are unnecessarily high, but more probably will result in some of the bidders ignoring the restrictive provisions entirely on the theory that "they can't possibly be enforced; so why include them in cost considerations?" And when the successful bidder begins to ignore requirements that appear unreasonable, the owner's representatives must either enforce the specifications, with wrangling and risk of lawsuit, or else waive the objectionable requirement—a dangerous precedent for the job, and certainly a concession that is unfair to the other bidders.

One difference between well-drawn plans and specifications for a drilled pier job and those for a more conventional construction project is that the drilled pier plans must be more flexible, for underground conditions must always be inferred from limited data, and the exact conditions that will be encountered at a site will usually remain unknown until the last pier hole is drilled. This means, at the very least, that some flexibility must be allowed in dimensions; there must be provision for field decisions to found piers deeper (or shallower) than planned, or to change shaft or bell sizes, or to place additional piers when required by deviations in pier location or plumbness. And in some geologic situations economy may dictate a change in pier type—or even a change to another type of foundation—as the work proceeds and the underground picture becomes clearer.

The plans, the specifications, and the contract documents should reflect the need for this field engineering. The needed flexibility will vary from a minimum for sites where the geologic formations are regular, uniform, and well known, to a maximum for sites where heavy loads or important structures are to be carried by irregular deposits or by decayed, fissured, cavernous or pinnacled rock. The project specifications should be consistent with what is known—and what is unknown—about site subsurface conditions.

Drilled pier plans and specifications are sometimes prepared by engineers or architects who are not thoroughly familiar with the problems and techniques of drilled pier construction, with local geology, with what is known about site conditions, or with local pier-drilling practice and equipment. This can lead to misunderstandings and complete disagreement between owner, engineer, and contractor, with consequent job delays and cost increases. Attempts by an inspector or engineer to enforce compliance with vague or inappropriate requirements will usually result in weakening of the authority of the engineer, relaxation of the project specifications, and ultimate claims for "extras."

Specifications should not be prepared on a "rush" basis. When drafted, they should be reviewed by the project geotechnical engineer and by a lawyer, with especial attention to clarity, completeness, consistency, and the presence of terms that could be misinterpreted.

It is neither prudent nor economical to use a standard set of specifications for all drilled pier jobs, changing only the pertinent dimensions and numbers. The specifications for each job should be written to provide for all contingencies that are predictable on the basis of available site information, and should allow as well for the possibility of unforeseen field changes. This means that each set of specifications should be carefully tailored to the job at hand. Pay items designated in the bid documents and contract should cover all probable work items, so as to allow the contractor to get a fair price for all his work without having to ask for "extras" or claim "changed conditions." Care should be taken to avoid unrealistic quantity estimates for any pay item that affects the bid total, for such errors are an invitation to unbalanced bids from contractors who recognize the error because of their local experience. Typical of this sort of error is an underestimate of a rock excavation pay item. For sites located in areas where ground conditions are erratic, the bid documents should state that quantity estimates are uncertain and should call for over and under unit prices, with a statement that bids will be rejected if they contain unit prices that are, in the judgment of the owner's representative, unbalanced.

The authors have seen disastrous cost increases where these precautions in the bidding documents had been neglected. Because of these possibilities, the bid and contract documents, as well as the plans and specifications, should be prepared or reviewed by a geotechnical engineer experienced in all phases of design and construction of drilled piers, and acquainted with site geologic conditions, before bids are advertised.

Bid documents should also require bidders to submit a list of projects in similar subsoil and groundwater conditions that they have completed successfully. For projects of special difficulty, or where proven capabilities and experience are essential, prequalifying prospective contractors and negotiation of contracts in lieu of competitive bidding is recommended to provide the owner with the best and least expensive construction.

As has been stated elsewhere, the designer of the structure is normally responsible for the production of redesigns made necessary by ground conditions which result in deviations of piers from position, alignment, or plumbness. The designer is usually designated in the contract documents as the structural engineer; or the terms "engineer of design," "engineer of record," or "architect/engineer" may be used. In the discussion that follows, and in the sample specification (Art. 7-3), the term "structural engineer" will be used for this agency, and the terms "geotechnical engineer" and "foundation inspector" will be used as defined in Chap. 5.

The scope and the limitations of the duties that go with these positions should be set forth clearly in the contracts between the owner (or his agent) and these several agencies. They may be included also in the construction contract, or may be incorporated therein by reference to the special contracts concerned. In either case, the contractor should be enabled to acquaint himself with the duties and limitations of the engineering/inspection agencies that he will have to deal with.

Contracts often require the contractor to furnish inspection services as part of his contract. The object is, of course, to enable the owner to include the cost of inspection as part of the construction cost when applying for a construction loan or for a mortgage. This practice is wrong; it is against the interests of both owner and lender. Any inspection agency *must* represent the interests of the owner. An inspection agency which is paid by the contractor can be subjected to conflict of interest, no matter how good its intentions or how firm its resolve. An inspection decision that costs the contractor money may be protested; disputes may develop; and the contractor, resentful at being required to pay for what he considers "bad" decisions, is in a position to express his resentment by holding back, or even refusing, payments to the inspection agency. This is not merely a hypothetical possibility. It happens on many contracts, and it benefits nobody. An inspection agency should be engaged by, report to, and be paid by the owner or one of his agents. Lending agencies would find it to their interest to recognize this situation and allow inclusion of inspection costs as part of construction costs.

The presentation of soil or rock profiles as part of the plans is a particularly risky procedure. Any such presentation will normally be prepared by a draftsman, usually without training in geology or soil and foundation engineering. Errors in wording or symbols are common. It is imperative that all subsurface data presented in the plans be checked *in every detail* by the geotechnical engineer, both to be sure that all data have been correctly presented and to be sure that nothing in the presentation might be taken as fact when it is no more than inference.

All of the subsurface information available to owner, architect, structural engineer, or geotechnical engineer should be made available to prospective bidders; and this should include the soil and rock samples from the subsurface exploration, preserved at natural moisture content and arranged so that they can be readily examined. The samples, of course, will have to be kept at some central location, such as the architect's office. The geotechnical report (or soil and foundation report), with complete boring logs, should be reproduced and made available to prospective bidders. *Making these documents available for examination in the architect's office is not sufficient. A contractor needs to be able to study them freely and at length while preparing his bid.*

7-2 A CHECKLIST FOR SPECIFICATIONS

To be complete, a specification for drilled pier construction should cover many items, both general and specific for the project covered. It is easy to overlook and fail to cover an item which can come up later as the basis for a question, or a dispute, or even a claim for an "extra." To help the reader check his specifications for completeness, the following checklist is offered. It should not be expected to cover all possible contingencies. New structural requirements, new construction techniques, variations in ground conditions—all these things require variations in specifications. Perhaps the use of the checklist will suggest some of these variant items.

SPECIFICATIONS CHECKLIST

7-2.1 General Conditions

(a) *Site*—Identify project and site, list physical restrictions such as overhead restrictions or obstructions, noise restrictions, access limits, underground obstructions, and utilities.

(b) *Subsurface*—List all subsurface investigations and reports, publications, or other sources of subsurface information available to contractors, including availability of soil samples and rock cores for examination.

(c) *Statement of Contractor's Responsibility* to familiarize himself with site and subsurface conditions, and that ignorance of conditions will not be accepted as a basis of claim for additional compensation. The authors believe it would be helpful if the bidder's representative were required to sign a statement that the geotechnical data furnished, including soil and rock samples or cores, have been examined.

(d) *Facilities Available to Contractor* during construction—power, water, sanitary, storage areas, etc.

(e) *Lines and Grades*—Locations of installed piers to be determined by a survey to be provided by _____ .

(f) *Permits*—List those required, those to be provided by owner, those to be provided by contractor.

(g) *Insurance*—List types and amounts required before contractor can start construction.

(h) *Structural Plans, Drawings, Specifications*—List these and state where they can be obtained.

(i) *Dates* for start and completion (defining extent of completion by that date), and sequence (if required) of operations. (When geological conditions are such that a change in foundation design is considered possible, it may be desirable to specify that the sequence of drilling will be designated by the geotechnical engineer.)

(j) *Geotechnical Engineer and/or Foundation Inspector*—Define their duties (both scope and limitations), their authority, and the lines of communication between them and the contractor and the owner.

(k) *Methods of Construction*—Omit or keep as nonrestrictive as practicable.

(l) *Deviations from Specifications*—State that they are permissible only if approved in writing by the geotechnical engineer.

(m) *Bid Items and Estimated Quantities*—List these, and list separately any contingency items for which unit prices are wanted but which do not enter into the bid total.

7-2.2 Special Conditions

(a) *Pier Alignment*—Maximum permissible variation of center of pier from design location (line and grade) and deviation from plumb, to be specified by structural engineer. (Common values are 1 to 3 in. from center, $1\frac{1}{2}$ in. out of plumb for first 10 ft, $\frac{1}{2}$ in. for each 10 ft of additional depth.) These tolerances should be kept as large as possible.

(b) *Batter Piers*—Suggested maximum allowable deviation of batter pier is 5 percent (in angle of inclination). (Note that while the angle with the vertical is easily controlled, horizontal direction is not.)

(c) *Permissible Increase* in nominal axial load resulting from deviation from location or plumbness—to be decided by the structural engineer for each instance.

(d) *Additional Construction* (including costs of engineering and redesign) required by deviation from tolerances and as specified by the structural engineer will be done at no additional cost to the owner.

(e) *Shaft and Bell Dimensions* should be at least as large as shown on the plans. Deviations should be permitted only with written permission of the structural engineer.

(f) *Excavation*—List limitations (if any) as to method or machinery.

Restrictive limitations increase costs and should not be specified unless necessary.

(g) *Disposal of Excavated Material*—Specify disposal area, or state that this is a responsibility of the contractor. Specify street or highway cleanup requirements.

(h) *Temporary Casing*—Specify, if required by law or by known geologic circumstances. (This refers to casing left in place until concrete has been placed or casing as protection during inspection or cleanout.) Provide that casing will be used whenever requested by foundation inspector. Specify outside diameter to be not less than nominal size of shaft, wall thickness sufficient to prevent crushing or deformation of casing by earth or water pressure, and that casing must be watertight. No payment to be made for casing abandoned by contractor.

(i) *Casing Left Permanently in Place* at the direction of the geotechnical engineer, or permanent casing required by the plans, should be a priced item in the bid. (Some governmental agencies and some local authorities require permanent casing of all pier holes.)

(j) *Bearing Stratum and Bottom Depth*—Specify and describe in geologic terms for rock, and use unified soil classification system [7.1] for soils. State where samples of the designated bearing material can be examined. Define sound rock, weathered rock, boulders, and cobbles, as required by expected or possible underground conditions. For example, very hard cemented layers, constituting hard rock for excavation purposes (ironstone, for example), sometimes occur in shallow soil formations where bedrock is hundreds of feet deep. If such are known to be possible, provision should be made for a suitable contingent pay item.

State that bearing stratum or strata will be identified by the foundation inspector, who will indicate to the contractor where the hole should be stopped or who may indicate, after inspection of the hole, that the hole should be taken deeper.

(k) *Belling or Underreaming*—Specify bell diameters; specify bell shape only if required for some special reason. [The authors consider all the usual bell shapes (Fig. 1-6) to be equally satisfactory in performance.]

(l) *Obstructions*—Define, and specify procedures to be followed in cases where obstructions are encountered. (This should be a contingent pay item, priced on a time basis, in the bid documents. Basis for payment should be clearly set forth.)

(m) *Dewatering*—Specify the maximum depth in inches of water that will be permitted in the bottom of a straight-shaft pier, or in the bottom of a bell or underream, at the time of placement of concrete. Specify procedure to be followed if dewatering proves difficult or impractical, and who is to decide that dewatering is not feasible. [Alternates: (a) seal off water by installation of casing; (b) grout casing at bottom if necessary, then drill through grout; (c) leave water in hole and place tremie concrete or pumped-in concrete in accordance with appropriate section of the project specifications.]

(n) *Bottom Cleanout*—Specify the maximum depth in inches (suggested: ¼ in.) of loose material, mud, or "slop" that will be allowed on the bottom when concrete is placed. Specify that a 2-in. layer of clean crushed stone, ½- to 1-in. size, be placed on the bottom if required by the geotechnical engineer.

(o) *Proof-testing* of rock, if needed, should be described and the contractor's duties defined. At the option of the structural engineer, tests may be required (a) for every pier hole, (b) for a fixed proportion of pier holes, or (c) at the option of the geotechnical engineer. (This should be a pay item.)

(p) *Load Tests*—If load tests are required, specify details of tests, including range of loads, load increments, length of time each increment is to be maintained; state the approximate number of tests that will be required, and locations; state that the tests will be performed by the contractor, but will be observed by the foundation inspector and/or the geotechnical engineer. (This must be a pay item.)

(q) *Safety Provisions*—List minimum equipment and personnel to be furnished by contractor, and state that all equipment must be available and assistance present during every descent into a hole. (Unless written instructions to the contrary are received from the geotechnical engineer, every pier hole should have protective casing installed before entry of any person into the hole. A fresh air supply, either an air hose from a compressor or a safety air tank to be worn by the person entering the hole, should be required before entry of any person into the hole.)

(r) *Measurement of Pier Locations*—Specify who is to locate holes with respect to their design positions (line and grade) immediately on completion of the drilling, and again on completion of the concreting.

(s) *Inspection of Pier Holes*, including measurement of depth, of plumbness, and of pier and bell diameters, will be done by the foundation inspector, who will be furnished by (name, address, and telephone number of firm or organization).

(*t*) *Approval Before Concrete Placement* will be required from the foundation inspector for every pier hole, and no concrete will be placed unless the foundation inspector is present.

(*u*) *Concreting*—Specify prompt placement; 7-day and 28-day strength; limits of slump; and special requirements for concrete to be used with reinforcing cages or to be placed by tremie, "elephant's trunk," or pumping.

(*v*) *Concrete Sampling*—Specify frequency of sampling, name of sampling and testing agency, and tests required. (For very important or heavily loaded structures, one test cylinder should be taken from every batch or truckload of concrete delivered to the site.)

(*w*) *Reinforcing Steel*—Positioning of cage must be approved by the sampling and testing agency before concrete is placed.

(*x*) *Vibration of Concrete in Pier Holes*—Permitted? Required? To what depth?

(*y*) *Continuous Placement of Concrete*—Specify continuous placement, but provide for procedure to be followed if placement has to be interrupted (see Art. 4-5).

(*z*) *Removal of Temporary Casing*—Specify that casing may be broken free of bottom seal by a 2- to 4-in. jerk, then lifted by a smooth, vertical pull; concrete level to be maintained high enough to prevent intrusion of loose soil or water into shaft; observation of concrete surface; vibrating puller if required.

(*aa*) *Records*—Specify that contractor will keep logs of pier holes and daily records; pay items will be measured by the foundation inspector, and the contractor's representative can be present to verify quantities when the measurements are made.

7-3 A SAMPLE SPECIFICATION

The sample specification which follows presents a form that the authors have found satisfactory and includes some items that are often omitted from specifications for this type of work. Like other "standard" or "sample" specifications, it should not be used without careful study to adapt it to the proposed construction. For example, straight-shaft piers in rock, depending partly or entirely on sidewall friction for support, may require special attention by the contractor to cleaning or roughening of the walls of the hole. Because the nature of this preparation may be governed by (1) the kind and condition of the rock penetrated and (2) the drilling equipment and techniques used, a specification provision for cleaning or roughening may be quite different for different projects.

The user should scrutinize the sample specification carefully, rewrite any paragraphs that do not fit his project exactly, delete any parts that are not applicable, add special provisions needed for his project, check the result against the checklist on the preceding pages, and submit his final draft to his geotechnical engineer and his attorney for comment.

A SAMPLE SPECIFICATION FOR DRILLED PIERS

7-3.1   General Conditions

(a) *Location:* The location of this project is _____

_____

The structure to be constructed is _____

_____

The following underground obstructions, hazards, and site restrictions are brought to the attention of the contractor: _____

_____

(b) *Subsurface Information:* The following soil and foundation investigations have been performed for this site: _____

_____

Soil samples and rock cores obtained during those investigations may be examined by the Contractor at the following location: _____

_____

and copies of the soil and foundation report (or geotechnical report) will be furnished on request.   It is the bidder's responsibility to familiarize himself with site and subsurface conditions before submitting his proposals.   Ignorance of conditions will not be accepted as a basis of a claim for additional compensation.

(c) *Facilities Available to the Contractor:* The following facilities are available for use by the Contractor, subject to the following conditions and restrictions: _____

_____

(d) *Lines and Grades:* Lines and grades and all surveys will be furnished by _____

_____

(e) *Permits:* The following permits will be required before construction will be permitted at the site: _____

_____

The following permits will be supplied by the Owner for the Contractor: _____

_____

The following permits must be obtained and supplied by the Contractor: _____

_____

(f) *Insurance:* The following insurance, in the specified amounts, must be obtained by the Contractor before construction is started: _____

_____

(g) *Structural Plans:* Plans, drawings, and specifications for this project may be obtained from _____

_____

(h) *Completion:* Construction on the project is to start on or before _____

_____

Construction and completion of all work for various parts of the project shall be completed by the following dates: _____

_____

A sequence of construction operations has been established for this project as follows: _____

_____

(i) *Duties of Other Persons or Agencies*
STRUCTURAL ENGINEER: The Structural Engineer must approve, in writing, any changes in pier construction which involve or require changes in structural design of elements above the piers, or any changes in foundation dimensions or type which may be recommended by the Geotechnical Engineer. The Structural Engineer will furnish to the Contractor, as promptly as practicable after receipt of data on deviations from specifications, designs for such supplementary construction and/or reconstruction as may, in the Structural Engineer's judgment, be necessary. The Structural Engineer may, at his discretion, waive redesign and reconstruction when computed stress increases produced by deviations from specifications are not more than _____ percent of design stresses. The Structural Engineer designated for this project is _____
CONSTRUCTION REVIEW: The agency responsible for engineering construction review of the general contract for the project is (the Structural Engineer) (_____). The _____ _____ Engineer will furnish lines and grades and initial locations for piers, will check pier locations before and after concrete is poured, and will advise the Contractor immediately of any deviations from designated pier locations that are shown by the surveys.
GEOTECHNICAL ENGINEER (alternates: Soil Engineer, Soil and Foundation Engineer): _____

_____

is designated as Geotechnical Engineer for the foundation construction phase of this project. The Geotechnical Engineer is responsible for all phases of inspection of foundation construction except for surveying and for the sampling and testing of concrete, and is responsible for inspection of reinforcing steel. He is responsible for the daily approval and certification of pay items in the foundation contract. The Foundation Inspectors report to him and are supervised by him. The Geotechnical Engineer does not direct the Contractor's men, but advises the Owner or his representative of any deviation of the work from the specifications or from the intent of the contract documents, and keeps the Contractor informed of all of his advisories to the Owner. The Geotechnical Engineer shall decide whether or not drilling operations have reached suitable bearing material, and the Contractor is directed to comply with his decisions in this regard. All items of advice or recommendation by the Geotechnical Engineer to the Owner will be recorded in the Geotechnical Engineer's daily reports. On completion of the job, the Geotechnical Engineer will submit a written report stating that in his opinion all piers have been installed in accordance with project specifications and contract documents, or listing exceptions.

FOUNDATION INSPECTOR: The Foundation Inspector(s) for this contract will be furnished by (_____) (the Geotechnical Engineer), and will report to and be supervised by the Geotechnical Engineer. (Alternate: The Geotechnical Engineer designated above will act also as Foundation Inspector.) The Foundation Inspector(s) will keep field notes of all pay items completed each day, of all advisories or recommendations to the Owner, and of job and weather conditions; and will keep a log of each pier hole drilled, recording soil or rock strata, water entry and flow, caving, sloughing, squeezing ground, drilling difficulties, casing insertion, bottom cleanout and water removal, depth and size of hole and underream, details of rebar placement and centering, concreting, casing withdrawal (including casing abandoned or casing left in place under the contract), and any other construction details covered in his instructions. He will submit a report of these observations at the end of each day. He will check the Contractor's pier logs and daily summary of pay items completed against his own records, and will indicate his approval by initialing these records. His inspection reports will be transmitted daily by a means that assures their reaching the Geotechnical Engineer at the beginning of the following work day, and any advisories that record unsatisfactory construction practices or rejected work will be reported to the Geotechnical Engineer by the Foundation Inspector(s), by personal contact or by telephone, immediately. [The Foundation Inspector(s) will check as-drilled and as-built pier locations by direct measurement from batter boards furnished by _____ and preserved by the Contractor.] [Alternate: The Foundation Inspector(s) will not check as-drilled and as-built pier locations, this being the responsibility of _____
_____.]

CONCRETE INSPECTION: _____

has been designated as the agency for testing concrete delivered to the site and placed in the foundations. An inspector from this agency will take at least four cylinder samples, two for 7-day test and two for 28-day test, from each morning's and each afternoon's pours;* and will test selected batches for slump and reject any batches of concrete which fail to meet slump specifications, or which contain deleterious materials or have had water added to the mix to maintain fluidity within one hour of time of placement.

The Concrete Inspector will also observe the placement of the concrete for compliance with specification requirements for avoidance of segregation; for flow of the concrete outward through the reinforcing steel cages (if any); and for continuity of the concrete column, "necking down" of the concrete column, or possible intrusion of soil, mud, or water into the column of concrete.†

(j) *Construction Methods:* It is not the intent of these specifications to unnecessarily restrict the Contractor in his construction methods, techniques, or equipment. However, methods, techniques, or equipment herein specified (if any) are considered necessary to provide adequate pier installation. Deviations from these techniques or equipment may be made if approved in writing by the Geotechnical Engineer.

If the construction procedures used are observed to cause loss of ground or produce other conditions detrimental to the construction or to other property or structures, the Contractor will be required to use other and more appropriate procedures or to demonstrate that by changing some aspect of the procedure in question he can complete the construction safely and without the objectionable effects.

The Contractor will be required to correct at his own expense any subsidence, structural damage, or other objectionable conditions caused by his operations.

(k) *Bid Items:* The following items are to be done for the unit prices and/or lump sums inserted in the tabulation below:

| | | | Unit prices | |
| Item | Base quantity | Lump sum | Over base | Under base |

(1) *Prequalification Experience, Equipment, Personnel:* All bidders must submit to the Owner or his representative (at least _____ days in advance of the opening of bids) (alternate: as part of his bid) an account of his experience in installing the specified type of drilled pier

---

* The authors recommend that a test cylinder be taken *from every truckload* where the pier is to carry very heavy loads or is a part of a very large, heavy, or important structure.

† This inspection, being one of the most critical in the entire drilled pier construction sequence, should be the duty of both the Concrete Inspector and the Foundation Inspector, who will file separate reports.

foundations, a list of his available equipment and personnel, and a list of his current and pending (bid) contracts.

7-3.2 Special Conditions

(a) *Pier Alignment and Dimensions:* The piers shall be installed as shown on the applicable design drawings and in accordance with these specifications. No pier shall be off center from its design location more than _____ in. at the top of the pier. No vertical pier shall be out of plumb more than _____ percent of its length. Batter piers shall not deviate more than _____ percent of their lengths from design inclination. All piers and shafts shall be at least as large in diameter as shown on the design drawings. Deviations from underream (bell) configurations may be made only with the prior written approval of the Geotechnical Engineer. The maximum allowable overload of any single pier, resulting from eccentricity of the pier or pier group with respect to the imposed load, shall be less than _____ percent.

If any of the above tolerances are exceeded, additional construction (including costs of engineering and redesign) as required by the Structural Engineer shall be paid for by the Contractor.

Notes to Structural Engineer: (1) Blanks above are to be filled in by Structural Engineer; (2) Tolerances should be as large as practicable; (3) See checklist for suggested tolerances.

(b) *Excavation:* Excavation of pier holes shall be performed by rotary drilling methods, using any practical methods and machinery approved by the Geotechnical Engineer. It is anticipated that a drilling-machine torque of _____ lb-ft may be required for advancing some of the pier holes to the assigned bearing level, and the Contractor shall have on the job suitable machinery and tools to meet this requirement.

(c) *Disposal of Spoil:* All material removed from the pier holes shall be removed from the ground around the casing before concrete placement is started and (hauled to _____) (disposed of as directed by the Geotechnical Engineer), and the Contractor shall be responsible for avoidance and cleanup of street spillage to the satisfaction of the local authorities.

(d) *Protective Casing:* Protective casing, at least as large in inside diameter as the nominal shaft size and of sufficient wall thickness to resist crushing by hydrostatic and earth pressures, shall be installed (in each pier hole) [when needed, in the judgment of the Foundation Inspector(s)] to prevent caving or fall-in.

Casing left in place at the decision of the Geotechnical Engineer is a pay item; that left in place by the Contractor without approval by the Geotechnical Engineer is not a pay item. Any casing found to be crushed or deformed by earth or water pressure shall be removed and replaced by suitable casing of heavier weight, which shall be left permanently in place as a pay item, but the crushed or deformed casing shall not be a pay item.

Alternates

(The decision as to whether or not protective casing will be left in place permanently will be made by the Geotechnical Engineer.)   (Permanent casing will be required for all pier holes.)

(e) *Obstructions:* For the purposes of this contract, an obstruction will be defined as any rock, cobble, or other object that cannot be removed or drilled through by normal drilling procedures and tools, using a drilling machine exerting a torque of _____ lb-ft. If at the Contractor's request the Geotechnical Engineer has given approval for trial use of a machine with smaller torque rating, no apparent obstruction shall be counted as a pay item unless the Contractor shall have brought in a machine of the above-indicated torque rating and shall have demonstrated to the satisfaction of the Geotechnical Engineer that the obstruction cannot be removed or broken up with that machine, using suitable drilling or chopping tools.

(f) *Water in the Excavation:* Any water above an average depth of _____ in. above the bottom of the excavation shall be pumped or removed before placement of concrete will be permitted.   In cases where water cannot be held below this level long enough for concrete to be placed in the normal manner, the Contractor may, with the approval of the Geotechnical Engineer, place concrete by one of the following methods: (*a*) use of a submersible pump in the bottom of a straight hole or in a sump excavated in the bottom of a bell, with concrete being placed to cover the intake pipe before the pump is lifted; (*b*) use of a tremie pipe or "elephant's trunk"; or (*c*) use of pumped-in concrete, discharging through a pipe below the water and below the surface of the concrete in the hole.

(g) *Underreamed (Belled) Piers:* Where underreamed piers are required, the Foundation Inspector will inspect and approve the bearing materials and depth, and will determine whether or not, in his judgment, the upper part of the underream would be in suitable stable material, before the belling process begins; and the Contractor shall drill deeper before starting the underream if advised by the Foundation Inspector that this is necessary.

(h) *Proof-testing of Rock:* In pier holes where proof-testing of rock at bearing level is specified (or is directed by the Geotechnical Engineer or his representative), the following testing procedures shall be followed to investigate the soundness of the supporting materials below the designated bearing level.   The equipment shall be a percussion drill (alternate: a rotary coring drill) capable of drilling a vertical hole approximately 2 in. in diameter to a depth below the bottom of the hole equal to twice the design diameter of the bearing surface or 4 ft, whichever is greater.   The drill shall be operated by a man experienced in its use, and the drilling shall be accomplished under a relatively uniform downward pressure in a manner such as to avoid binding of the drill bit.   The size of the bit may be reduced as the hole is deepened, if necessary to avoid binding.   The time required for each successive 6 in. of penetration will be recorded by the Foundation Inspector, who will also record the depth and extent of any fractured zone, seam, or void as reported to him by the drill operator. The Foundation Inspector will probe each test hole with a "feeler" rod, logging any fractured or soft zones or voids encountered, and will determine from this inspection and from the information received from the drill operator whether suitable bearing surface has been reached or whether the hole will have to be deepened.*

(i) *Safety Provisions:* The Contractor shall have at the job site all equipment and materials needed to provide safe descent into pier holes for the Foundation Inspector and any other personnel who may need to enter the holes, and shall provide personnel to observe any person at all times when he is in a hole. Equipment furnished by the Contractor and kept immediately adjacent to the hole at all times when any person is in the hole shall include hoist, ladder, or other means for descent into the hole and hoist line with safety hook for rescue purposes. In addition, the Contractor shall have on the site protective casing [see Sec. (d) above], air supply, electric light (safety lamp) on safety cord long enough to reach bottom of hole, and submersible pump. The use in the hole of a pump driven by an internal combustion motor will not be permitted. Electrically driven pumps, if used in the hole, shall be of the explosion-proof type.

Any person entering a pier hole shall be required to wear a safety hat, a safety rescue harness, and an air tank of approved type unless the Contractor provides an air line from a compressor. (This equipment, when worn by the Foundation Inspector, is part of his kit and is not furnished by the Contractor.)

(j) *Inspection:* All pier holes shall be inspected by the Geotechnical Engineer or the Foundation Inspector (1) at the time of drilling, to make sure the assigned bearing stratum has been reached; and (2) prior to the placement of concrete, to make sure the hole is in proper condition for concreting. Sufficient time shall be provided to permit inspection of the pier hole and all dimensions to be made by the Foundation Inspector. The location of the pier as drilled and its plumbness shall be determined before concrete is placed. If the location of the pier does not meet the required tolerances, concrete shall not be placed in the pier until permission is received from the Structural Engineer. Sufficient equipment and personnel shall be supplied by the Contractor to the satisfaction of the Geotechnical Engineer to permit proper inspection of all drilled pier excavations.

---

* The method of proof-testing described above has been found satisfactory for use in the weathered gneisses and schists of the Philadelphia area. In other geologic formations, more appropriate test methods may be specified by the Geotechnical Engineer. The number and selection of piers to be proof-tested should be designated in advance by the Geotechnical Engineer and approved by the Structural Engineer (who should keep in mind that the cost of such testing is little, while the increase in assurance of safety is great).

(k) *Reinforcing Steel:* Reinforcement steel shall be installed as shown on the design drawings. All steel shall be free of rust, mud, or any deleterious material which would hinder bonding of concrete and steel. Reinforcement cages shall be straight and shall conform to the design dimensions. Adequate provisions shall be made to ensure that the reinforcement steel will remain in place throughout placement of concrete and that specified concrete cover for the reinforcement steel is attained and maintained. The use of precast concrete spacer blocks for this purpose is recommended.

(l) *Concrete:* Promptly after approval of the excavated pier hole has been given by the Foundation Inspector, concrete shall be placed in a manner that will not cause segregation of the particles or permit infiltration of water or any other occurrence which would tend to decrease the strength of the concrete or the capacity of the finished pier. No concrete shall be placed in a drilled pier excavation unless the Concrete Inspector and the Geotechnical Engineer or the Foundation Inspector are present. All concrete shall have a 28-day strength of at least _____ psi and a slump between _____ and _____ in. Concrete placed by tremie (or "elephant's trunk") through water shall have one extra sack of cement per cu yd, up to a level 10 ft above initial water level. Concrete may be placed in a dry hole by dumping in free-drop from the surface, provided that a hopper or other device approved by the Geotechnical Engineer is used to force the concrete to drop straight down without hitting the sides of the holes or the reinforcement before striking the bottom. If battered piers are used, the concrete shall be placed with an "elephant's trunk" or tremie pipe that extends to the bottom of the pier. Concrete shall be placed continuously for the entire depth of the pier. In case placement has to be interrupted for emergency reasons, dowels will be installed as specified by the Structural Engineer or a 2- by 4-in. cross key shall be formed in the surface, laitance will be removed, the surface will be roughened and cleaned to assure bond, and the roughened surface will be slushed with a 1:1 cement grout just before more concrete is placed. Concrete placed in drilled pier excavations shall be vibrated as follows: _____

_____

(m) *Pulling the Casing:* An initial "jerk" of 2 to 4 in. will be allowed to start the lift; thereafter, while being removed from the pier hole, the casing must be kept plumb and must be pulled with a smooth, vertical motion, without jerks. Vibration of the casing during pulling is (required) (approved). A positive head of not less than 3 ft of concrete above the pressure at the bottom of the casing from the outside water table shall be established before the casing is lifted, and shall be maintained as the casing is pulled.

Where cutoff elevation is below ground level, the contractor will be required to maintain protective casing to the ground surface if needed to prevent detrimental caving or intrusion of shallow soils into the shaft.

Dowels will be placed and positioned after casing has been pulled and top surface of concrete has been established.

(n) *Records:* The Contractor will keep a log of each pier showing: (*a*) pier number; (*b*) pier location; (*c*) depth drilled through overburden; (*d*) depth drilled in bearing stratum; (*e*) elevation of ground surface; (*f*) top elevation of concrete; (*g*) top elevation and length of casing; (*h*) diameter of shaft; (*i*) diameter and type of bell; (*j*) spacing, depth, and condition of sidewall grooves, if required; (*k*) estimated inflow of water, source, and depth in bottom of hole when concrete is placed; (*l*) description of bearing stratum; and (*m*) pumping required.

The Foundation Inspector will confer with the Contractor's representative at the end of each day's work, and (if the Foundation Inspector is in agreement with the logs) both parties will sign the Contractor's logs of work completed on that day, and a summary statement of pay items will be completed. These records will be kept and signed in duplicate, one copy for the Contractor and one for the Owner or his representative. In the event of disagreement as to work completed, both claims will be entered on the summary sheet and the disagreement will be called to the attention of the Geotechnical Engineer promptly.

REFERENCE

7.1 Teng, W. C., "Foundation Design," Prentice-Hall, Inc., Englewood Cliffs, N.J., 1962.

CHAPTER EIGHT

Minimizing Construction Claims— Excessive Costs—Liabilities

Foundation construction is probably more subject to cost overruns, claims for "extras," and budgeting uncertainties than any other phase of building construction. The deeper the foundations, the more troublesome these cost uncertainties are likely to become; and drilled pier foundations are probably more subject to unanticipated costs than most other types of deep foundations. The reasons for this are inherent in subsurface variations: soil and rock conditions can be significantly variable within very short distances; the knowledge of subsurface conditions available to the contracting parties is always limited and imperfect; and the techniques of pier drilling and concreting are sometimes more sensitive to variations in soil, rock, and groundwater conditions—as well as to the effects of bad weather—than, for example, pile-driving operations. When unanticipated costs or claims develop, they may involve large sums of money, sometimes more than doubling the cost of the subsurface phase of a project. They may delay (or even prevent) the completion of construction; and they can be very damaging to the economics of the project, to the contractor's reputation, and to the engineer's or architect's relations with his client. The threat of major cost increases may require redesign, or reduction of the scope of a project, or may involve difficult revision of financing arrangements. If cost overruns occur frequently in the practice of an engineer or architect—or in the performance of a contractor, or in construction, generally, in a particularly area or region—they can adversely affect the availability of insurance coverage.

Such cost increases and claims cannot be eliminated completely, but they can be minimized, if their causes are understood and their avoidance is included in planning and in contracting.

8-1 CHANGED CONDITIONS

This term is used to designate conditions which affect the cost of construction and which are different from the conditions described in the plans and specifications, the contract, or any of the supporting data or information made available to the contractor. The term is not usually intended to imply that conditions have changed from what they once were but rather that they are different from those originally represented. "Changed conditions" clauses are put into specifications with the hope that bidders will feel that their risk of loss due to adverse subsurface conditions is covered and that they can reduce or eliminate the contingency allowance in their bids. Most commonly, "changed conditions" are described as in the following example taken from a set of specifications for a major project:

> *Changed conditions*—The Contractor shall notify the Engineer in writing of the following conditions, hereinafter called changed conditions, promptly upon their discovery and before they are disturbed:
> (1) Subsurface or latent physical conditions at the site of the work differing materially from those represented in this contract; and
> (2) Unknown physical conditions at the site of the work of an unusual nature differing materially from those ordinarily encountered and generally recognized as inherent in work of the character provided for in this contract.

"Changed conditions" in this sense, if proved, often constitute the basis for settlement or court awards to contractors. It is to the advantage of everyone concerned that construction conditions be known as completely as practicable before plans are prepared or bids are taken.

To this end, subsurface exploration programs should be planned with both the requirements of the structure and the probable subsurface conditions of the site in mind, and the man in charge of exploration should expect to modify his plans, if necessary, as exploration proceeds. It does not make sense to lay out in advance an array of borings to preassigned depths, without regard to what may be encountered below the surface when the borings are made. If a "changed conditions" claim is not to be encouraged, the exploration program must be adequate, both in horizontal coverage and in depth of borings, and must be reported in detail and in terms that contractors can understand. Moreover, samples of the formations penetrated must be preserved at natural moisture content for examination by prospective bidders. A discussion of exploration for drilled pier projects will be found in Chap. 6.

The best prevention for "changed conditions" claims is to make certain that you "tell it as it is" to the prospective bidder on your pier job. We can state from experience that some of the most common grounds for "changed conditions" claims on drilled pier construction lie in the areas listed below, when the contractor was uninformed or misinformed:

1. Type of rock or obstruction encountered (limestone, sandstone, shale, boulders, cobbles, "hardpan")
2. Accuracy of depth of rock layers
3. Variation between hard and soft layers
4. Stability of unconsolidated material to be drilled
5. Presence of squeezing or caving layers
6. Depth to water table and probable seasonal variations
7. Permeability of water-bearing layers
8. Presence of artesian pressures
9. Structure and hardness of rock to be drilled (joints, fracturing, dip, foliation, thickness of weathered zone)
10. Frequency, size, and shape of cobbles or boulders encountered
11. Unexpectedly difficult access
12. Excessive overdrilling (for various reasons, but sometimes due to inadequate—or inexpert—inspection)

The contractor must estimate his costs from what the plans, specifications, and other documents furnished tell him or do not tell him. These must be accurate and unambiguous. Most subsurface conditions tend to be heterogeneous rather than uniform, and usually subsurface reports and specifications are based on relatively few small-diameter borings for jobs requiring many drilled piers. It is important that in addition to presentation of the logs of the borings, the soil and foundation report should include results of observation of geologic conditions at and at some distance surrounding the site, in order to give the contractor as complete a picture as possible of what to expect. For instance, the fact that small borings do not encounter cobbles or boulders does not necessarily mean they are not there, especially in a type of sedimentary formation that commonly includes cobbles. Also it is not enough to say merely that cobbles and boulders are present. Their size and shape and frequency should be predicted if at all possible.

It would be prudent for the engineer to include some large-diameter borings in the geotechnical investigation of a site where there appears to be a possibility of extensive use of drilled piers. Although this may increase the cost of the investigation, it also may avoid delays and troublesome claims during or after construction.

If water is not present in a test boring, it is not sufficient to omit a water table symbol on the log. There should be a specific statement that it was absent and at what time of the year, and whether there might be water during another season.

These are only a few illustrations of the sort of "changed conditions" pitfalls that can be encountered. The engineer who is trying to avoid "changed conditions" claims must be willing to tell the bidder that certain conditions are unknown to him, if this is true, and must remember to put the responsibility upon the bidder to find the answer or make his own interpretation before he bids. And—very important—he must give the bidder ample time and opportunity to make his own study.

A last point of advice to both engineer and contractor on a large pier-drilling operation, especially a troublesome job, is that good documentation and record-keeping are most important. When difficulties are encountered by a contractor, the contractor's superintendent should make a *complete* record in his diary; and the inspector or the geotechnical engineer should record the nature of the difficulty, the contractor's methods in handling the problem, the type and condition of the equipment being used, the estimated time lost, and his opinion as to the cause of the difficulty. Photographic records of the situation can be of real value in settling claims before they become formal, or in arriving at an equitable settlement in case claims subsequently arise.

Although design changes during construction are sometimes considered to *constitute* "changed conditions," they are more commonly used as *evidence* of "changed conditions." Most design changes are covered under a special paragraph in specifications and generally make negotiation of revised payment necessary if the change is significant.

8-2 EXTRAS

Cost increases due to "extras"—items not anticipated or priced in the bid documents or contract—are almost inevitable on a large project, but they can be minimized by planning and by alertness on the part of all concerned as the construction proceeds. If there is a possibility of a new cost item being incurred, although it is not expected, it should be prepared for by inclusion of unit prices or a cost-plus price for the item in the bid forms or contract, but without forming part of the bid total. An example would be a unit price for permanent casing in a contract in which it is not expected that permanent casing will be required; or unit prices for various ranges of quantity of rock excavation in a site where rock is geologically possible but was not found in the exploration program. Such bid unit prices should be scrutinized by the geotechnical engineer before the contract is awarded. If they are "out of line," they should be renegotiated before the bid is accepted.

Sometimes such unit prices are not included in the bid documents but are written into the contract, usually for the specific purpose of avoiding unbalanced bids. These unit prices should receive as careful consideration by the geotechnical engineer as those in the bid forms; in fact, they would probably be more difficult to renegotiate later, or to revise in court action, than would unbalanced prices originating with the bidder.

8-3   CAUTION IN SELECTING FOUNDATION TYPE

The sensitivity of the cost of drilled pier foundations to geologic details, to weather, and to equipment type and condition has been mentioned in several places in the preceding chapters, and is discussed in some detail in Chap. 2. The claims for "extras" and "changed conditions" that are possible when a drilled pier job "goes sour" can be very large.

In circumstances where prolonged bad weather could interfere substantially with drilled pier construction, it might be necessary to change from piers to a pile foundation in order to keep to construction schedules. If this is a possibility, it should be recognized in advance and provided for as a contingency in the general contract, with pile foundation designs ready for use if necessary and contingent bid prices to match. If a change of this sort has to be made during construction without advance preparation, it is almost sure to result in substantial increases in cost and time.

The authors have seen several instances where a pier-drilling contractor has approached an owner, sometimes after a contract for another type of foundation has been let, and persuaded him to change from a pile foundation to drilled piers by promising a substantial saving in cost. The geotechnical engineer should be consulted before any such move is consummated, and if the owner goes ahead with a change against his advice, he should withdraw from the job at once. In one instance where the authors' firm stayed on the job after such a change, the contractor valiantly struggled with adverse soil conditions to complete the job, but took so long to do it that the increased costs of engineering construction review and inspection substantially exceeded the contract saving. The reader is invited to guess who felt the weight of the owner's dissatisfaction.

8-4 SUPERVISION, INSPECTION, AND CONSTRUCTION COSTS

The competence and performance of the geotechnical engineer and the foundation inspectors can go far toward minimizing claims of "changed conditions." The geotechnical engineer should be assigned the responsibility for planning and adjusting the exploration program, and for its reporting. He should review plans, specifications, and bid documents as they relate to presentation of subsurface information and to drilled pier operations; and he and the foundation inspectors should keep in mind at all times that their operations and reports are the basis for the owner's, the structural engineer's, and the architect's day-to-day understanding of the contractor's progress in relation to subsurface conditions. During construction the geotechnical engineer should be immediately aware of any deviations—or apparent deviations—of ground conditions from those anticipated from the earlier exploration and postulated in the contract documents; and he should make decisions promptly, without delaying the contractor's operations and, on the other hand, without allowing any unnecessary work that the owner will have to pay for. To this end, both geotechnical engineer and foundation inspector should know the assigned allowable bearing pressures for the various expected strata, and also the basis for these allowances—whether governed by code, local experience, conservative estimates, or field or laboratory test results. Where local geologic variations result in a question whether or not the stratum reached is the same as—or equivalent to—an assigned bearing layer, a decision must be made whether to stop or to continue drilling until a more competent—or, perhaps, a more familiar—formation is reached. This decision should be made by someone who understands both the support requirements of the structure and the structural properties and behavior of soil (or rock) in all variations that might be encountered. This is the geotechnical engineer's duty; only to a limited extent can such a decision be delegated to an inspector, and then only to the extent that possible geologic variations can be preidentified for him. If this type of decision is left to the contractor, you may be sure that he will not take a chance: he will go deeper "just to be sure" and the owner's budget will suffer, sometimes quite unnecessarily. This is not meant as a criticism of drilling contractors. Their ideal task is to drill a clean, dry hole, of dimensions, location, and depth as directed; and to place in it steel and concrete in such a way as to ensure the continuity of load transmission. If a contractor can do this and keep to the construction schedule, his job is well performed. He should not be asked—or permitted—to do the work of the geotechnical engineer.

Two points that require special care in inspection of drilled pier holes are cited below. Inadequate performance of inspection in either case has resulted in major increases in foundation cost.

One aspect of inspection that requires special care is the decision on where to stop when drilling in soft or weathered rock. The load-supporting capacity of even soft rock is usually much higher than is generally supposed. Current design practice is moving in the direction of taking advantage of this property, with factors of safety against failure due to overload approaching 5, 4, or even 3 instead of 10 or 20 as formerly used (though not recognized). (Refer to Chap. 3 for discussion of allowable loads on—and in—rock.) At the same time, pier-drilling machines are developing in power and cutting ability, so that a rock that might have represented "refusal" to the largest machine available locally last year might be readily drillable with today's equipment. The foundation inspector must be prepared to recognize and designate the assigned bearing stratum or bottom depth from inspection of the cuttings *plus descent into the hole* and examination of sides and bottom—*not from difficulty or ease of drilling.* Unless geotechnical engineer and foundation inspector are alert to this aspect, pier holes may be taken to entirely unnecessary depths when more than normally powerful equipment is used.

The precise opposite is true when the drilling equipment is not powerful enough, or the cutting bits or teeth are not suitable: "refusal" may be encountered before assigned depth is reached, and the contractor may have to go to jackhammer work ("dental work") to take the hole down. The cost of rock removal by this means may increase by $200 or $300 per cu yd (or even more), and a claim for "extras" or "changed conditions" is almost inevitable. This points up the importance of specifying minimum size or power of equipment in certain cases (Chap. 7), and the importance of making at least some of the preliminary exploration borings with large-diameter augers (Chap. 6).

Another typical circumstance requiring special care in inspection is the presence of seams of silt or fine sand in a formation that is nominally a stiff or hard clay. Minor seams of silt below the water table may have an alarming appearance to an inspector, yet be of no significance as far as the load-bearing ability of the formation, either in end-bearing or sidewall shear, is concerned. In geologic circumstances where this condition can occur, inspectors need special instructions regarding its significance with respect to their assignment, and geotechnical engineers need to keep in close touch with the inspection. It is reported that in the Chicago area, where silt seams are common in the Wisconsin and Illinoisan till, there was a period when overruns in drilled pier costs were very common because the foundation inspectors employed were using a pocket-size instrument for testing the shearing strength of the clay down in the hole, and did not recognize that the instruments had no legitimate application to the silt layers that were encountered.

The increased cost of construction that can result from inadequate or improper engineering supervision and inspection is not the only excessive cost that may be incurred. The prime objectives of these services are to obtain piers with good bearing in competent earth or rock and to obtain piers which are composed of competent, continuous concrete from bottom to top. Failure to accomplish either of these objectives can result in structural failure or disastrous reconstruction costs. The magnitude of these risks is illustrated by the following quotation from *Engineering News Record* (September 29, 1966, p. 15):

> At 500 N. Michigan Avenue (Chicago), a 25-story reinforced concrete office-apartment skyscraper, the top segments of the defective caissons are being removed and replaced.
>
> On the 25-story building, cracking in the lower flat-plate concrete floor slabs led to the discovery of a major defect in a 4-ft-dia column caisson. The top of the caisson had settled about $1\frac{3}{4}$ in. when construction reached the 23rd floor. Core borings in the caisson . . . led to the discovery of a clay-filled void in the concrete of the caisson shaft, about 45 feet below grade.

The example is not cited as an instance of inspection failure. The authors have no knowledge of the cause of the imperfections in these piers. It is cited to illustrate the possible magnitude of the damage which could conceivably follow when technical supervision and inspection have been inadequate or inexpert—or absent. The sky's the limit!

8-5 WHEN A DRILLED PIER JOB "GOES SOUR"

Trouble on a drilled pier construction job usually takes the form of impeded progress, or failure of the pier contractor to keep to schedule. Possible reasons, as has been indicated above, are many. A few that the authors have encountered:

Prolonged bad weather
Subsurface conditions different from those expected
Foundation type or design unsuited to ground conditions
Inadequate equipment
Inadequate or incompetent inspection
Inadequate experience of contractor's men
"Quick" or flowing silt, resulting in loss of ground
Excessive water inflow (often from gravel or cobble layers)
Unexpected rock or hard, pinnacled, or broken rock

Most of the discussion in the preceding paragraphs of this chapter has been concerned with avoiding delays due to any of these reasons. Of course, these reasons for delay should never be allowed to develop; but they do sometimes develop, and the question then arises: What to do about it?

When prolonged bad weather causes delays on a drilled pier job, the delays in subsequent construction may dictate a change to another type of foundation. As mentioned earlier, pile-driving rigs can sometimes operate under weather conditions which shut down pier-drilling operations. A change of this sort would be expensive, but might be necessary, for example, to allow the structure to meet occupancy commitments.

Subsurface conditions different from those expected should, ideally, never be encountered, but the subsurface is full of surprises. When such a condition does become evident, there should be a prompt meeting between owner, designer, and supervising engineer; and the geotechnical engineer should be a party to the conference and should bring all his experience and expertise to bear in helping decide what changes should and can be made. If the foundation design is not right for the site, this should be recognized and corrected as quickly as possible—regardless of whose fault it is.

If the contractor's equipment proves inadequate for the job, or if his men are not skilled at their jobs, it may be necessary to bring in another contractor. It should be recognized that the deposed contractor will probably claim "changed conditions"; and this points up the necessity for special care in documenting the new contractor's methods, equipment, and progress; and for very careful logging of some of the pier holes that he drills, for comparison with the information originally available to the first contractor. Ordinarily, detailed logs of pier holes are not made. In this instance, such logs might be of substantial help in litigation. .

Excessive water inflow may be very difficult to control, especially when the water is carried in a layer of cobbles or broken rock over sound bedrock. When the sealing off of water becomes too difficult or time-consuming, it may be best to let the water rise to equilibrium in this hole and place the concrete by tremie or by pumping. This adds to the cost of the concrete, but may be a more acceptable solution than struggling with the bottom seal until it is finally tight.

Unexpected rock conditions that require hand-tool excavation can be very troublesome. At the present time and in most areas, jackhammer work in the hole is usually the only solution. A roller-bit tool, patterned after oil well drilling tools but using air instead of water or mud to cool the bit and remove the cuttings, is under development, and has been reported to be very promising. It is to be hoped that it will come into general use in the next few years.

When a site proves geologically unsuited to drilled piers, as demonstrated by the contractor's first efforts, another type of foundation should be designed and a suitable contract let as quickly as practicable. In such a case there should be careful consideration of the possibility that the contractor's equipment or techniques may be at fault. But a site that is really unsuited to drilled pier construction can produce no end of trouble if efforts to overcome all its difficulties are persisted in. The authors recommend prompt decision and redesign in such a case.

8-6 THE PROFESSIONAL LIABILITY PROBLEM*

Engineers and architects have become increasingly—and painfully— aware of the professional liability problem during the late 1950s and through the 1960s. Geotechnical engineers have been particularly affected. At the time of the writing of this book, professional liability insurance is practically nonexistent for geotechnical engineers, except with an extremely high deductible (the amount which the engineer self-insures). For other engineers and architects the professional liability insurance market is very poor. Only one or two companies are offering coverage to this latter group, the coverage is becoming extremely expensive, deductibles keep rising, and there is an increasing number of exclusions in the policies.

It has been pointed out that drilled pier construction is particularly sensitive to claims. The authors therefore believe that the professional liability problem merits detailed discussion in this book.

8-7 THE EVOLUTION OF THE LIABILITY PROBLEM

Until about the mid-1950s professional liability suits against engineers and architects were rare. The huge increase in such suits is probably a by-product of our increasingly complex civilization. For one thing, along with the large increase in population, there was a tremendous increase in the total amount of construction. This almost automatically implies that a lot of new people, many probably lacking in experience, were becoming involved in the industry. Also new specialties developed, and new types of engineers and subcontractors were appearing

---

* Most of the following material was developed either jointly or concurrently by the authors and Risk Analysis & Research Corporation, Edward B. Howell, President. We want to thank Risk Analysis & Research Corporation for permission to use the portions developed by them.

on the scene. The pace of the individual project was stepped up because of new equipment, and also because increasing costs made it imperative to have the project completed and functioning as rapidly as possible.

As of 1971 it is common for an owner—once a decision to go ahead with construction has been made—to require that a plant go "on stream" (be in production) so as to meet a very specific time schedule. For example, a brewery would have to be able to meet the summer beer demand, a shopping center would have to meet a fall opening date, or an automobile assembly plant would have to be in operation by the time new models are being introduced. If such schedules are not met, huge losses result. Under these circumstances there is tremendous pressure to complete the project within the allotted time. Possible delays due to weather, or to ground conditions different from those shown in the plans, or to inept inspection or engineering supervision can be very harmful to the owner under these conditions.

These developments have had a serious effect on communications between all of the parties involved. In the days when there were only one or two engineers, an architect, one or two contractors, and the owner on a project, all the parties knew one another. If something went wrong, it was comparatively easy to get together and work out a solution. With the increased number of people involved and the stepped-up pace, it is no longer possible—or at least it is extremely difficult—for all the people involved to know and properly communicate with one another.

The result of all these developments has been a large increase in the number of suits against engineers and architects. This situation caught engineers, architects (and other professionals), and their insurance companies unprepared. It became apparent that large judgments were possible, and an avalanche of suits followed. The United States also has entered an age of "consumerism," in which people expect every article to be "guaranteed." As a consequence there has been an effort by some attorneys to extend the doctrine of implied guarantees (or warranties) of products to cover services provided by engineers and architects.

The design profession is now learning that it is unwise to take on responsibilities for which they have not contracted and are not getting a fee. Engineers have to be extremely careful not to create "pockets of exposure" in their reports, plans, and specifications—and particularly in their construction review and inspection, when these functions are part of the contract. It is important for the geotechnical engineer recommending or inspecting drilled piers to exercise extreme care. As a first step, it would be prudent for him to insist on following through with the field inspection on any project where he has submitted a soil or foundation report calling for the use of drilled piers. It is also essential to have properly trained foundation inspectors on the job, and to have the geotechnical engineer immediately available at all times. Many problems can be avoided if good communications are maintained with the other parties involved, and if any problem that does occur is attacked immediately.

8-8 LIABILITY TO THIRD PARTIES

It is well to keep in mind all parties that might be harmed by any error or omission—real or alleged—on the part of the engineer or architect. One usually thinks of the owner, the party with whom the architect, engineer, and contractor have contracted, as the only one to whom they are responsible. Unfortunately, this is far from true. Many buildings, for example, are built by one party and immediately sold to another. The land for which an investigation was made may be sold, and a third party can build on it without the knowledge of the geotechnical engineer who made the report. Yet this third party may place real or alleged reliance on the opinions expressed in the report. Or, in a different situation, a contractor—or a drilling subcontractor—can sue the owner for extra charges and the owner can in turn try to collect from the engineer or architect by means of a cross complaint.

If drilled piers are used as shoring, a claim may be filed by an adjacent property owner, alleging that the material removed during construction of the piers removed lateral support of his building and caused settlement. The architect or structural engineer also could be sued by any of these parties, and he in turn could file a cross complaint against the geotechnical engineer. The insurers for any of the unknown third parties, or lenders financing the construction to which damage allegedly occurred, also can file cross complaints against any of the various parties involved—including the geotechnical engineer. Or a suit may be instituted on behalf of an "injured workman," someone employed by the contractor, whose recourse against his employer is restricted to the limited recovery of workmen's compensation, while his potential recovery against others is unlimited.

8-9 THE COST OF PROFESSIONAL LIABILITY CLAIMS

As indicated in the previous paragraphs, suits or claims concerned with construction may—and generally do—involve many parties. For example, an injured third party, when filing a suit for alleged damage caused by drilled pier construction on an adjacent property, could name the owner, the contractor, the drilling subcontractor, the architect, the structural engineer, the geotechnical engineer, and probably others. This multiplicity of defendants makes the defense of such claims extremely expensive for all concerned. Each of the parties listed in the example above may have one or more attorneys representing him. (For example, a defendant may use his corporate attorney to protect the deductible portion of the claim, while another attorney may represent the portion covered by insurance.) As a result ten or more attorneys may be attending seven or more sets of depositions. A court reporter, together with numerous copies of transcriptions of the testimony given, is involved in each deposition. Many of the parties may hire their own "outside experts." If the case goes to court, because of the number of parties participating in the suit, a lengthy trial often results.

Because of such complications, the cost of defense accounts for the major portion of the liability losses suffered by engineers and architects in all claims, including those involving drilled piers. Losses due to judgments have in most cases been minor, with the exception of judgments in favor of an "injured workman." (Several of these latter cases, however, have been reversed by an appeals court.) It is not to be implied that no major judgments have been rendered against engineers or architects for alleged errors or omissions connected with drilled pier installations or other projects. There have been large judgments rendered, and probably a fairly substantial number of small ones. The principal cost of professional liability claims against engineers and architects, however, is connected with defense.

8-10   LOSS PREVENTION PRACTICES

The best way to avoid a claim, of course, is to have no problems. Drilled pier construction, unfortunately, tends to be somewhat problem-prone. It is important then for the geotechnical engineer to keep three significant objectives in mind when working on a project involving drilled piers: (1) to plan and carry out a comprehensive investigation, thoroughly covering all possible problem areas, both in the investigation and in the report; (2) to make certain that the owner is aware of the limitations on the degree of certainty of the findings reported, as well as the effects of weather, adequacy of equipment, and contractor's experience; and (3) to insist on carrying out the responsibilities of construction review and inspection with properly trained personnel who are able to maintain effective communications with the other parties on the job.

Many suggestions for accomplishing the above objectives have been discussed previously. In making the field investigation, geology, as well as the use of geophysical and other remote sensing methods (when applicable), may be helpful. The owner should be made aware of the fact (and it should be clearly spelled out for the benefit of the drilling contractors submitting bids) that a boring log represents only the soils *at the location of the boring, and only at the time the boring was made.* (For example, a water table may vary with time.)

Geotechnical engineering is not an exact science. Even though a most comprehensive investigation may be undertaken, the actual soils sampled can represent only a very small fraction of the total amount of material that will be encountered during actual construction. Natural soil deposits can be extremely variable, and extrapolations between borings should be classified only as "best estimates." The exact conditions are not fully known until the excavations have been made and the foundation completed.

The owner and the engineer or architect preparing the contract plans should be advised not to put boring logs on the plans. The bidder should be advised as to where he can examine the report describing the complete geotechnical investigation, including boring logs and samples. Not only can errors or omissions be made in copying logs, but the report is necessary for proper interpretation of the boring logs.

It should be stressed to the owner that the experience and reliability of the drilling subcontractor, together with the adequacy of his equipment, are extremely important. If the job has a strict time schedule, the owner should be made aware of the problems that can be caused by adverse weather, and contingency plans should be developed for emergency use. Training of field technicians—both in technical aspects and in good communications—cannot be overemphasized, nor can the primary necessity of doing careful inspection and construction review.

8-11 CONTRACT LANGUAGE

The engineer and architect should carefully explain to the owner what should be included in the contract in order to best protect all parties against "changed conditions" or "extras" (discussed previously). It is imperative that language be included in the contract between the owner and the contractor, holding the engineer and architect harmless from injuries or damage arising out of the contractor's operations and not due to the *active* negligence of the engineer or architect (the injured workman claim, for example). To accomplish this protection, it must be made clear to the owner, and the contract language should specify, that any "supervision" or "inspection" performed by the engineer or architect is only for the purpose of obtaining compliance with the plans and specifications, and that the engineer and architect *do not* and are not expected to have any supervision whatsoever of the contractor's operations. In preparing the contract, it also should be explained to the owner that job safety is the sole responsibility of the contractor, and the contract should state this fact specifically.

Courts tend to strictly limit the application of "hold harmless" clauses as being against the public interest. The language described above, however, in the opinion of most attorneys, is valid and will protect the engineer and architect.

8-12 LIMITATION OF LIABILITY

Limitation of liability for direct damages to the owner arising out of errors or omission by the engineer is possible if included in the contract between the engineer and the owner. It has to be reasonable under the circumstances; for example, it has to be evident that limitation of liability is not forced on the owner. It could be spelled out, for example, that the owner had a choice of accepting a limitation *or* paying a higher fee to cover additional insurance costs. Laws governing limitation of liability, like other laws, vary from state to state. The architect or engineer should have his attorney familiarize himself with the pertinent details in states in which such contract language may be available for inclusion.

8-13 STATUTES OF LIMITATIONS

Statutes of limitations can be helpful in some cases, but again they vary considerably from state to state. The courts in some states have a tendency, however, to be very liberal in their interpretation of the statutes, particularly with respect to innocent (or presumably innocent) third parties. In many states, for example, the courts have held that the statutes do not start to run against third persons from the time of the completion of the project, but rather from the time of discovery of any latent damage. Furthermore, in the case of continuing damage (for example, settlement of adjacent structures, which may continue for months or even years), it has been held that the statutes do not start to run as long as damage continues to occur.

As an example, an existing building adjacent to a heavy structure on drilled piers might settle due to the additional load imposed on the underlying strata. The condition may be latent (present but not readily visible) for several years before the damage progresses to the extent that it becomes apparent. It could continue for several more years. The interpretation of the courts in some states may be that the statutes of limitations do not start to run against the owner of the adjacent building until the damage has stopped. In such a case the injured party could file a suit perhaps 15 years after the completion of the structure, naming all parties who conceivably could have contributed to the damage.

8-14 IMPLIED WARRANTY

An earlier mention was made of the tendency on the part of the public (fostered by decisions reached by courts, consumer suits, and new laws designed to protect the consumer) to expect that everything they purchase is warranted to be satisfactory for the purpose intended. There actually have been attempts to extend this "implied warranty" to persons rendering services, such as engineers, in several cases in California. Generally, however, the courts have applied the so-called "prudent man" rule. This requires that an engineer giving recommendations for drilled piers (for example) exercise the same degree of judgment and care that any reasonable and prudent engineer would exercise under the same circumstances—that is, in the same or similar area and at the time that the services were provided. In view of the present tendency to protect the consumer, good judgment would seem to demand that nonwarranty provisions be included in proposals, reports, specifications, contracts, or any other documents pertaining to an engineering project.

8-15 SUMMARY

In summary, it is prudent, when drilled piers are called for in a design, for the engineer and architect to exercise careful judgment both in the engineering and in the contract language; to make full disclosure of the degree of reliance that should be placed on the geotechnical information, the possible effects of weather, and the influence of the contractor and his equipment; and to follow through with competent construction review and inspection.

The engineer or architect should not rely on his own knowledge when legal interpretation is involved, but for his own protection should consult with an attorney. This is especially important insofar as protective clauses in contracts are concerned. As previously mentioned, laws governing such clauses (as well as limitation of liability, nonwarranty provisions, and statutes of limitations) vary considerably from one state to another, and the advice of competent attorneys is essential to engineers and architects in protecting themselves against liability claims.

As previously stated, the best way to avoid a claim is to have no problems. Unfortunately, once someone sustains any damage, everyone who conceivably could have contributed to the alleged damage can be named in a suit. A good plaintiff's attorney is required to do this to protect his client. (Naming multiple defendants also is an inexpensive way to gather information by means of depositions.) Once named, the engineer or architect must defend the suit, even though he may have made no contribution whatsoever to the damage. In view of these circumstances, it is not enough for the individual engineer or architect to exercise extreme care; in addition, every attempt should be made to get all the parties involved in the project together at its inception, particularly when drilled piers are involved. All possible problem areas should be discussed, so that everyone concerned with the project has a complete understanding of them. If a friendly attitude and good rapport can be established and maintained, problems that do arise generally can be resolved without the need for a claim to be filed.

Characteristics of Auger Machines (Manufacturers' Data, 1970)

Maximum hole diameters and depths given in the following tabulations are manufacturers' estimates for ideal drilling conditions, and should be considered as limiting values rather than continuous operating capacities. Deeper and larger holes can often be drilled by use of drill stem extensions, special reamers, etc.

Characteristics of Calweld Drills (1970)

Crane-mounted rigs, to use either drilling bucket or auger. Maximum hole diameters given are for drilling bucket with reamer; maximum depths are for longest kelly (deeper holes can be drilled).

Model	Max. hole Diameter, ft	Max. hole Depth, ft	Max. continuous torque, lb-ft	Max. downward force, lb
Series 200	20	Depends	350,000	22,000
Series 175	15	on crane	176,000	plus
Series 150	10	and kelley	122,000	weight
Series 100	8		93,000	of kelly
Series 75	7		75,000	and tools
Series 50			52,000	

Characteristics of Calweld Drills (1970) *(Continued)*

Truck- and crawler-mounted rigs, for bucket only.   Maximum hole diameters are for bucket with reamer.

Model	Max. hole Diameter, ft	Depth, ft
1500A	17	210
1000A	12	130
SPC-500(Crawler)	11	85
250B	11	85
200B	10	85
190B	9	70
150B	7	70
100B	6	70

Characteristics of SS 8487 Crane Attachment Caisson Drill (1970)

(W-J Sales Engineering Co., Inc., Houston, Tex.)

Model	Max. hole Diameter, ft	Depth, ft	Max. continuous torque, lb-ft	Max. downward force, lb	Max. batter
SS 8487 Single unit	10	150	150,000	*	45°
SS 8487 Double unit	15	175	300,000	....	45°
SS 8487 Double with adapter	20	200+	800,000	....	45°

* Maximum downward force depends on type of mounting and weight of unit and crane.

Characteristics of Earthdrill Machines (1970)

Model	Mounting	Max. continuous torque, lb-ft	Max. downward force, lb	Max. Hole Diameter, ft	Depth, ft	Kelly bar size, in.	Max. batter
5200	Truck	91,000	38,000	10	250	4¼	1:3
52	Truck	91,000	26,000	10	200	to 7	1:3
52 HMM*	Truck	98,000	30,000	10	225	6/4½/3½†	1:3
45	Truck	90,000	24,000	8	200		1:3
45 HMM*	Truck	63,000	29,000	10	225	6/4½/3½	1:3
	Crane	to 500,000	29,000	8	130	to 7	1:3
520	Truck	67,000	28,000	8	250	to 7	1:3
36	Truck	69,000	22,000	8	200	to 7	1:3
32	Truck	69,000	20,000	6	200	to 7	1:3

* All-hydraulic model.
† Indicates triple-telescoping kelly.

Characteristics of Hughes Diggers (1970)

Model	Mount-ing	Max. continuous torque, lb-ft	Max. continuous force, lb	Max. hole Diam-eter, ft	Depth, ft	Kelly bar size, in.
				Heavy duty		
CLLDH	Crane	102,000 @ 5 rpm	15,000	10	100–120	11/7*
LLDH	Truck	59,000 @ 12 rpm	50,000	10	80–120	7/5¼
LDH	Truck	21,000 @ 23 rpm	37,000	8	25–100	6/4¼
CLDH	Crane	38,000 @ 10 rpm	10,000	8	50–100	6/4¼
				Medium duty		
MF	Truck	10,000 @ 32 rpm	23,000	6	30–60	5¼/4/3½
				Light duty		
CDH	Truck	13,500 @ 15 rpm	12,000	5	12–20	3
EF	Truck	5,500 @ 65 rpm	10,000	4	12–20	2½

* Double figure 11/7 indicates telescoping kelly, inner shaft 7 in. square solid steel, outer 11 in. square (outside measurement).

Characteristics of Texoma Holediggers (1970)

Model	Mounting	Max. continuous torque, lb-ft	Max. continuous down force, lb	Max. hole Diam-eter, ft	Depth, ft	Kelly bar size, in.
DMB 100	Truck*	44,000	45,000	8	100	7/5½†
600	Truck	23,500	26,000	6	35	3
500	Truck	23,500	26,000	6	40	3
330	Truck	21,000	26,000	6	40	3
300	Truck*	5,500	10,000	4	20	2½
270	Truck	16,000	16,000	4½	20	2½
254	Truck	14,000	16,000	4	20	2¼

* All-hydraulic diggers—all other diggers are available with hydrostatic drive. All diggers except DMB 100 are available as crawler mounts.

† Indicates telescoping kelly; outside 7 in. square; inside 5½ in. diameter.

Characteristics of Watson Caisson Drills (1970)

Model	Mounting	Max. continuous torque, lb-ft	Max. downward force, lb	Max. hole		Kelly bar size, in.	Max. batter
				Diameter, ft	Depth, ft		
Heavy duty							
6000-CA	Crane	200,000	Variable	15	Variable	6–12	30°
3000	Crane carrier or crawler	103,000	30,000	10	120	4–8	90°
5000-CA	Crane	103,000	Variable	10	Variable	4–8	30°
1000-TM	Truck	76,000	15,000	6	30	4	90°
1000-CM	Crawler	76,000	15,000	6	30	4	90°
Light duty							
800	Truck	13,500	13,000	4	20	4½	90°*

* Any angle between 0 and 90° from the horizontal is attainable, as the boom lies horizontally when boomed down.

Stabilizing Action of Drilling Mud

For those who have not actually used drilling mud, an explanation of what it is and how it works might be useful. Drilling mud is a thin slurry of clay in water. We say a "thin" slurry because it is not viscous to the degree that is suggested by the word "mud." Its viscosity is somewhat higher than that of water, but not conspicuously so as far as its appearance and "feel" are concerned—that is, as long as it is kept moving. The drilling mud slurry is readily pumpable.

If it is allowed to stand, in a few minutes a good drilling mud slurry will gel, assuming a "body" and consistency something like that of clabbered milk, and this structure keeps the clay from settling out of suspension. This slurry is thixotropic—i.e., it will become fluid again if stirred vigorously, and will gel again if allowed to stand again. Its density is, of course, somewhat greater than that of water because the specific gravity of the clay particles is somewhere in the range of 2.7 to 2.8, but the difference is, in the usual application, small; a typical drilling mud slurry might have a bulk specific gravity of 1.05 to 1.10, or a unit weight of 65 to 69 lb per cu ft. Heavier muds can be prepared by the addition of heavy minerals, for special purposes.

The clay minerals most used commercially for drilling mud are bentonite, for most applications, and attapulgite, or a mixture of attapulgite and illite, for applications where salt or other minerals react with bentonite to inhibit its swelling properties or to make it flocculate and settle out of suspension. However, many natural clay soils will form, at least temporarily, a usable drilling "mud" slurry when dispersed in water; and it is sometimes possible to prepare the slurry by mixing the soil being drilled with water during the drilling process and keeping this slurry agitated to keep the clay in suspension, since it will usually not gel as a commercial drilling mud does. A hole that is stabilized in this manner is called a "self-mudded" hole.

Before discussing the way a clay slurry acts to stabilize the walls of a bore hole, we need to consider the forces that tend to make a hole cave.

Above the water table, a hole through a cohesive soil tends to stand open because of the shearing strength of the soil; and because of "arching" effects that tend to develop the frictional strength of the soil as the walls of the hole try to close in, a very little cohesion is sufficient to keep the walls from collapsing. But if the soil is cohesionless—sand or gravel, for instance—cohesive forces are absent; the fall of one particle releases two or three more; and soon the entire hole has collapsed, leaving only a small crater at the top. Note that the collapse is progressive: unsupported particles fall in, leaving the next exposed particles unsupported, and this action is then repeated.

Below the water table, the same forces act, plus another set: the differential hydraulic forces set up by water moving either into or out of the hole. If water enters the hole, the forces set up are toward the hole, and unless the cohesive strength of the soil is sufficient to resist these forces, soil sloughs into the hole. After the water rises to groundwater level within the boring, the hydraulic imbalance disappears and the sloughing tendency is the same as that described above for dry soil—except that the weight of the soil grains is now buoyed weight rather than dry weight (a reduction of about 60 percent). And if the hole is filled with water from an outside source so that the level in the hole is above groundwater level, then water seeps *out of* the hole and the seepage forces on the soil grains are *outward*—i.e., in a direction tending to stabilize the walls of the hole.

So the key to keeping an uncased boring from caving is to keep it full of water—while it is being drilled and afterward, as long as you want it to stay open.

If water will do, then why use a drilling mud?

The slurry is heavier than water, and therefore exerts a greater stabilizing pressure.

Water from the slurry escapes slowly into the pervious soil, depositing a thin layer of clay on the walls of the hole. This clay layer is very impervious. Because the clay particles cannot move far into the voids of a soil—even a coarse sand—without becoming trapped and clogging the voids, the impervious layer thus formed (often referred to as a "filter skin") is very thin and concentrates the stabilizing pressure exactly at the surface of the wall—i.e., just where it is needed.

During the drilling process, the slurry, because of its higher density and higher viscosity, keeps cuttings in suspension far better than water does.

And because of the filter skin's very low permeability, the slurry loses water only very slowly, making it easy to maintain drilling fluid level, even in a boring through dry, coarse sand.

Moreover, drilling mud exerts a stabilizing pressure even in a boring through a very soft clay, which is so impervious itself that there is no loss of water into it from the slurry, and therefore no filter skin builds up. In this case, the hydrostatic pressure of the slurry balances, or partly balances, the lateral pressure of the soft clay; and usually the slurry pressure can be maintained high enough to keep the hole from "squeezing in."

"Mudding-in" Techniques

In using a clay slurry for keeping a pier hole open during drilling, visual observation and experience are generally used, rather than density and viscosity testing, for control of the slurry. The clay used may be a commercial product, or in some cases, the clay present in the soil being drilled. The technique is to add enough water and (when necessary) enough clay to mix with the soil being drilled and form a "mud" that is viscous enough to hold gravel or larger stone particles temporarily in suspension; heavy enough to stabilize the walls of the hole; and fluid enough to permit operation of the drilling auger, bucket, or other tools.

In a clean sand, this may require one 100-lb sack of good-grade commercial bentonite for every 4 or 5 cu yd (3 or 4 cu m) of material to be excavated. Drilling is performed in a loose sand by using an earth auger (see Fig. 4-17, Mudding Auger) and adding water and dry clay as the hole is deepened. The bit is lifted, rotated and "churned" to mix the water and clay into a uniform slurry. (Where mixing equipment is available, work is expedited by premixing the water and clay into a thin slurry before it is dumped into the hole.) Sufficient water must be supplied to keep the slurry thin enough for the auger not to become stuck under the heavy load, to fall freely through the processed slurry, and to come out freely with a load of cuttings when necessary. When the potentially unstable formation has been completely penetrated and an impervious or stable stratum has been reached, or the assigned depth has been reached, casing can be lowered to the bottom through the slurry and seated by driving or rotating. If the casing is sealed into an impervious and stable formation, the slurry can be pumped out or removed by a bailing bucket on the drill kelly.

When the hole is to be drilled in a predominantly silt and clay formation (often with potentially caving layers of sand), it can be drilled with a drilling bucket below a clay slurry. As the hole is deepened, more slurry is added (or mixed in the hole), with the slurry level kept always at or near the top of the hole. The drilling bucket should have a 4- to 6-in. relief passage to allow the slurry to pass downward as a load of cuttings is lifted; otherwise lifting the bucket will "suck in" the walls of the hole.

When beds of heavy gravel or boulders are encountered, the slurry may have to be thickened to help the drill bucket pick up the loose stones; otherwise they may tend to push or roll, and not be picked up by the bucket.

Information Sources

The concerns listed here were very helpful to the editor and authors in supplying catalogs, photographs, information, counsel, and advice.

PIER- ("CAISSON"-) DRILLING CONTRACTORS

The list that follows is not meant to be comprehensive. There are many capable pier-drilling contractors who were not consulted. We suggest that persons looking for a local contractor in this field consult the classified section of their telephone directory. Try the following categories:

Contractors—Foundation, —Drilled Piers, —Caissons
Drilling Contractors
Foundation Drilling
Caissons

(It would be helpful if telephone companies would adopt a uniform listing system for their classified directory sections.)

A. H. Beck Foundation Company, Inc., 5123 Blanco, San Antonio, Texas 78216. Tel. 512-342-5261.

Caisson Corporation, 6200 Gross Point Rd., Niles, Illinois 60648. Tel. 312-774-0166.

Case International Company, Box 40, Roselle, Illinois 60172. Tel. 312-625-1250.

George F. Casey Company, 7222 East Slauson Ave., Los Angeles, California 90022. Tel. 213-723-1201.

Drilling Service Company, 5121 No. Lindbergh Blvd., Bridgeton, Missouri 63042. Tel. 314-731-1111.

Farmer Foundation Company, 6818 Long Drive, Houston, Texas 77017. Tel. 713-643-2609.

Girdler Foundation and Exploration Company, Box 927, Clearwater, Florida 33517. Tel. 813-531-1455.

Layne Minnesota Company, 3147 California St. NE, Minneapolis, Minnesota 55418. Tel. 612-781-9553.

McKinney Drilling Company, Box 8, Nacogdoches, Texas 75961. Tel. 713-564-8373.

Meredith Drilling Company, Inc., 945 So. Huron St., Denver, Colorado 80223. Tel. 303-722-3253.

The Millgard Corporation, Box 2004, Livonia, Michigan 48150. Tel. 313-425-8550.

P & Z Company, Inc., 230 Harbor Way, South San Francisco, California 94080. Tel. 415-761-1850.

Raby Hillside Drilling, Inc., 8927 Ellis Ave., Los Angeles, California 90034. Tel. 213-837-2212.

Reliance Drilling, Inc., York, Pennsylvania.

N. L. Schutte Foundation Drilling, Inc., 1208 S. Industrial Blvd., Dallas, Texas 75207. Tel. 214-747-8525.

Smith & Brennan Pile Co., Inc.-Wabash Drilling Co., 110 Angelica St., St. Louis, Missouri 63147. Tel. 314-421-2640.

Watson Foundation Co., Inc., Box 11006, Fort Worth, Texas 76110. Tel. 817-927-8486.

Western Caissons Limited, Box 1725, Saskatoon, Saskatchewan, Canada. Tel. 343-1627.

Organizations

Association of Drilled Shaft Contractors, Box 47482, Dallas, Texas 75247

EQUIPMENT MANUFACTURERS AND DEALERS

Equipment catalogs from the concerns listed below were consulted, and the editor consulted representatives of nearly all of those listed—sometimes repeatedly. All were very patient and helpful.

Addresses and telephone numbers are included in the listings because these suppliers are sometimes difficult to locate. As far as the editor could find out, there is no central listing of *pier-drilling machines,* or *caisson drilling machines,* or *earth auger machines.* One would expect to find them in *Thomas' Register of American Manufacturers,* but with one or two exceptions they are not there.

Acker Drill Company, Inc., Box 830, Scranton, Pennsylvania 18501. Tel. 717-586-2061.

Calweld Division of Smith Industries International, Inc., Box 2875, Santa Fe Springs, California 90670. Tel. 213-723-0881.

Earthdrill, Inc., Division of George F. Casey Co., 7222 East Slauson Ave., Los Angeles, California 90022. Tel. 213-685-7369/213-721-4888.

Hughes Tool Company, Box 2539, Houston, Texas 77001. Tel. 713-926-3101.

Rotary Oil Tool Company, 8655 Whitaker Ave., Buena Park, California 90620. Tel. 714-526-4621.

Texoma, Inc., Box 998, Sherman, Texas 75090. Tel. 214-893-6371.

Watson Manufacturing Company, Box 11006, Fort Worth, Texas 76110. Tel. 817-927-8486.

W-J Sales Engineering Co., Inc., Box 36275, Houston, Texas 77036. Tel. 713-497-6793.

Foreign

Sir Robert McAlpine & Sons, Ltd., 80 Park Lane, London W1, England. Tel. 01-629-8234.

Drilling Mud Manufacturers' and Dealers' Literature

Baroid Division, National Lead Company, P. O. Box 1675, Houston, Texas 77001: "Drilling Mud—Aquagel," Sec. 2100, Baroid Drilling Mud Reference Manual; also "Quik-Gel" (a catalog information bulletin); also bulletins on bentonite for various purposes. [Source of montmorillonite-based drilling muds, baroid (heavy) muds, and various other drilling muds and chemicals for the drilling industry.]

Floridin Company, 3390 Peachtree Rd., Atlanta, Georgia 30305: "Florigel H-Y High-yield Salt Gel. Basic Data." (Source of attapulgite-type drilling mud.)

Johnson Division, Universal Oil Products Company, 315 N. Pierce St., St. Paul, Minnesota 55104: "Revert" (a sales bulletin); also reprints on Revert from the Johnson Drillers' Journal, issues of September–October 1965, September–October 1966, November–December 1969 (available from Johnson Division).

Safety Equipment

Mine Safety Appliances Company, 1100 Globe Ave., Mountainside, New Jersey 07092. Tel. 201-232-3490.

Pi-pellets (Bentonite Pellets)

Piezometer Supply Company, 700 Whittier St., Bronx, New York 10474.

Standards of
the Association of
Drilled Shaft Contractors
Box 47482, Dallas, Texas 75247

Item 1 Classifications for Drilled
Shafts (Caissons)

Type A. Not cased, not reinforced.
Type B. Not cased, reinforced.
Type C. Temporary casing, not reinforced.
Type D. Temporary casing with liner; not reinforced.
Type E. Temporary casing, reinforced.
Type F. Temporary casing with liner; reinforced.
Type G. Permanent casing, not reinforced.
Type H. Permanent casing, reinforced.

Underreamed shafts—Add the letter U to the classification type of shaft.
Battered shafts—Add the letter B to the classification type of shaft.
Submarine shafts—Add the letter S to the classification type of shaft.

NOTE: All types include concrete as specified.

Item 2 Diameters for Drilled Shafts (Caissons)

Shaft size nominal diameter, in.	Types A-B	Types C-D-E-F-G-H		
	Diameter of drilled hole, in.	Outside diameter of casing, in.	Diameter of drilled hole	
			For casing, in.	Extended thru casing, in.
12	12	Never cased		
14	14	Never cased		
16	16	Never cased		
18	18	18	20	16
20	20	20	22	18
24	24	24	26	22
26	26	26	28	24
30	30	30	32	28
36	36	36	38	34
42	42	42	44	40
48	48	48	50	46
54	54	54	56	52
60	60	60	62	58
66	66	66	68	63
72	72	72	74	69
78	78	78	80	75
84	84	84	86	81
90	90	90	92	87
96	96	96	98	93

Item 3 Underreamed or Belled Shafts (Caissons)

A. Maximum base diameter for sizes 12 through 24 in. of types A and B drilled shafts is three times the diameter of the drilled hole listed in item 2.

For all other sizes and types of drilled shafts, the maximum base diameter is three times the listed diameter of drilled hole extended through casing.

B. The fully extended bell angle is 45°.

C. The standard toe height for the bell is 3 in. for drilled shaft sizes 18 to 42 in. and 6 in. for sizes 48 to 96 in.

Item 4 Casings for Drilled Shafts (Caissons)

A. Casings are regular grade steel line pipe, produced by electric seam welding, butt welding, or spiral welding.

B. As listed in item 2, the outside diameter of casing for each size of drilled shaft is the same as the nominal shaft diameter.

C. Available wall thickness for each standard casing is as follows:

Standard casings, outside diameter, in.	Standard available wall thickness range, in.
18 to 26	Min. $\frac{1}{4}$; $\frac{9}{32}$; $\frac{5}{16}$; $\frac{3}{8}$ max.
30 to 36	Min. $\frac{5}{16}$; $\frac{3}{8}$; $\frac{7}{16}$ max.
42 to 60	Min. $\frac{3}{8}$; $\frac{7}{16}$; $\frac{1}{2}$ max.
66 to 96	Min. $\frac{13}{32}$; $\frac{7}{16}$; $\frac{9}{16}$; $\frac{3}{4}$ max.

Item 5 Tolerances for Drilled Shafts (Caissons)

A. Tolerances on the outside diameter and other dimensions of casings are the standard API tolerances applicable to regular steel line pipe.
B. All drilled hole diameters listed in item 2 are subject to normal drilling tolerance of $\pm\frac{1}{2}$ in.
C. All shafts shall be bored plumb to a tolerance of $1\frac{1}{2}$ in. for depths up to and including 10 ft plus an additional tolerance of 1 percent of the depth in excess of the first 10 ft.

Item 6 Reinforcing Steel Fabrication Standards for Deep Drilled Shafts

Purpose

To form a reinforcing cage to meet the engineering requirements as a structural element to retain its proportion throughout the placing of concrete and extraction of the casing from the drilled shaft.

Physical Characteristics

Longitudinal rebars per engineers requirements to be of the largest practical size, with #11 bars being the maximum. A minimum number of eight each longitudinal bars to permit a minimum of four in the lowest section, where longitudinal reinforcing bars are reduced.

Where spiral caging is required in the upper portion of the shaft, banding will begin at the bottom of the spiral and be spaced to the bottom.

Banding Characteristics

Banding material is $\frac{3}{8}$-in. $\times$ 2-in. strapping used through 42 in. diameter and $\frac{1}{2}$- $\times$ 4-in. strapping used through 96 in. diameter. Material is formed as a hoop, with the ID approximating the OD of the reinforcing longitudinal bars. These hoops are fabricated to the reinforcing stringers via fillet weld to each longitudinal bar for the full length of contact on one side. The hoop will have a minimum of 2 ft overlap of the ends and is welded across each end. Bands will begin at the bottom of the spiral and be equally spaced to 12 in. from the bottom. Spacing will not exceed 48 in.

A minimum of 2-in. concrete cover requirement will control maximum OD of reinforcing cage. Where casing is being utilized, reinforcing steel will be a maximum of 4 in. smaller OD than the ID of the casing. The apertures formed by rebar cage will be not less than 6 in. vertical and 3 in. horizontal.

Spaces in the form of a segment of a circle made from reinforcing smooth wire minimum section of #3 will be used to maintain concentric alignment.

They are to be tack welded in place to the longitudinal bars to project 2 in. outside the reinforcing cage OD. Spacers will be in four places at each elevation, spaced maximum of 10 ft on center below 50 ft and a maximum of 15 ft.

Reinforcing steel cages thus prepared may be handled on the job site by any method not including deforming or torsional or bending deflection.

Splicing of reinforcing steel is a standard rebar lap and fixed by tying or tack welding of this lap. Welding will be a minimum of 2 in. each weld—one at each end of the lap and one in the middle of the lap on one side (inside or out).

Steel will be placed firmly on the bottom of the excavated shaft or supported free from the bottom to 12 in. to permit a 12-in. subsidence without deformation.

Conventional and Empirical
Design for Bearing Capacity

As considered in Art. 3-11, piers drilled both in soil and in rock are often proportioned by "rule of thumb" methods which have little or no theoretical basis. In some cases the maximum allowable bearing pressures are specified, usually as "presumptive bearing capacity," in local building codes. Too often these figures are used in design without regard to differences in quality, conditions, or structure of what is nominally a specified, identifiable soil or rock material, but which may be subject to variations in real strength of the order of 10 to 1—or even more.

Sometimes laboratory or field testing to determine foundation material design parameters is not feasible. Sometimes such testing is not economical because it is possible to assume overconservative design parameters and still produce a safe foundation at less cost (and faster) than the cost of a testing program and the development of a more sophisticated design. And sometimes the foundation materials are uniform enough, and their properties are well enough known, that appropriate, safe, and fairly economical bearing pressures can be assigned on the basis of identification of the bearing material and an accurate survey of its structure and condition *in situ*.

Table E-1 on the pages that follow presents a range of design values commonly used to proportion drilled piers in some of the major cities of the United States. These values were reported by practicing foundation engineers in the areas covered. They are usually derived from actual experience and are often correlated with readily derived indices such as soil properties, seismic velocity measurements, diamond core recovery (RQD), standard penetration tests in bore holes, and pocket penetrometer and torvane tests.

These figures are not recommended for use as design parameters. They may, however, be of some use in indicating probable ranges of possible safe design pressures on certain materials and formations, and they may suggest possible ranges of allowable pressures on other materials or formations. They are also of interest in pointing up the differences in design approach between different regions—differences resulting principally from regional differences in the construction problems involved in reaching competent bearing materials.

Locale and source of information	Bearing material	Column load range, kips	Dia. range, ft — Bell	Dia. range, ft — Shaft or socket	Shaft or socket length/diameter range	End-bearing range, ksf	Sidewall shear range, ksf	Remarks
Greater Boston, Mass.	Marine clay (Boston blue clay)—varies from hard yellow clay to medium blue clay	80–300	4–10	2.7–4.0	15–40	2–10	zero	Typical machine-drilled and belled caissons, known locally as "Gow Caissons." Belling is generally performed in organic silt (organic silty clay) or in peat which directly overlies clay or s.-g. stratum. Concrete shaft is never cased. However, metal casing commonly used during installation. Bottoms are cleaned up by hand and visually inspected.
Information from: Haley & Aldrich, Inc. Cambridge, Mass.	Outwash sand and gravel (Loose to very dense)	100–250	4–8	2.7–3.5	15–35	4–10	zero	
	Shale or slate (Cambridge argillite)	1500–3000	No bell	2.5–3.0	100–150	100–150	15 in rock socket only	Drilled-in caissons
One Project	Shale (Cambridge argillite)	15,000	22	14±	65–70	60	zero	The Boston Company Building (40-story building completed in 1970). Four caissons—belled in rock by hand mining methods.
Chicago Area	Dolomitic limestone	1000+	—	3+	………	100	None	
	Dolomitic limestone	1000+	—	3+	1:1	200		With reinforcing from the permanent casing the loading can be increased to 200 ksf
Information from: Testing Services Corporation, Wheaton, Ill.	Hardpan	200+	4+	2.5 min	………	Up to 20	Usually none	
	Hard silty clay	100+	4 min	2.5 min	………	8–16	Usually none	
	Tough to very tough silty clay	20+	4 min	2.5 min	………	4–8	Usually none	
Dallas and Northern Texas	Residual clays	30–750	3–10	………	Up to 50	5–12	1–2	Sidewall shear values given can be used either alone or combined with end-bearing (for straight shafts only). Straight shafts are more commonly used in Austin chalk than belled footings.
	Alluvial clays	30–150	3–8	………	………	2–5	NA	
	Unweathered Austin "chalk"	30–2000	………	1–6	………	30–100	2–6	
Information from: National Soil Services, Inc., Dallas	Eagleford shale	30–2000	3–12	1–6	………	12–40	1–4	
	Taylor marl	30–1000	3–12	1–6	………	8–20	NA	

Source	Formation							Remarks
Denver, Eastern Colorado, and Southern Wyoming.	Claystone (Arapahoe formation)	30–200	5–20	1–10	Up to 20	15–30*	1.5–3**	Minimum deadload to resist uplift caused by wetting of overburden soils or rocks should be 50 to 100% of end-bearing. End-bearing for piers on surface of formation is about ½ of that for pier drilled well into it.
	Claystone & sandstone (Denver formation)	200–8600	5–20	1–10	Up to 20	40–60*	2–6**	
	Sandstone (Fox Hills formation)	30–6000	5–20	1–10	Up to 20	20–60*	2–6**	
	Claystone (Pierre formation)	30–6000	5–20	1–10	Up to 20	20–60*	2–6**	Minimum deadload to resist uplift caused by wetting of overburden soils or rocks should be 100% of end-bearing.
Information from: Woodward-Clevenger & Associates, Inc., Denver	Claystone & sandstone (Laramie formation)	30–2000	5–20	1–10	Up to 20	15–30*	1.5–3**	
	Claystone & sandstone (Chugwater formation)	30–2000	5–20	1–10	Up to 20	15–30*	1.5–3*	
	Clay	30–200	5–10	2–10	Up to 20	4–8	0.4–0.8** 0.6–1.0	
	Dense sand	30–200	N.A.	2–10	Up to 20	5–12	0.5–1.2** 0.7–1.7	

* Shear values for use in combination with end-bearing values in preceding column, for straight shafts only.
** These end-bearing values are for piers drilled well into the formation ("rock socket") and are used in combination with the sidewall shear values shown. End-bearing values for piers bearing on the surface of the formation are about half these values.

Source	Formation							Remarks
Detroit and Vicinity Information from: Halpert, Neyer & Assoc. Farmington, Michigan and Soil & Materials Engineers, Inc., Detroit, Michigan	Glacial till (Hardpan)	100–6600	3–15	2.5–8	Up to 50	8–50	None	Sidewall shear usually neglected in belled caissons.
	Dolomitic limestone	100–7000	2.5–5	2.5–5	Up to 50	20–200	None in overburden	Bond stress in rock socket taken as 50 to 250 psi (150 psi average.)
	Overconsolidated silty clay (not Hardpan)	30–1000	3–8	2.5–4	Up to 20	3–20	None	
Kansas City Area Information from: Woodward-McMaster & Associates, Inc., Kansas City, Mo.	Limestone	50–3000	to 8	2.5 min	Shaft* 4–20 Socket none	10–120	Not used	The limestone cannot be drilled without coring. End-bearing only is used; bells are formed in overburden. Design stress is based on limestone condition, thickness, and underlying materials. Shale is not used when weathered. Settlement potential may be critical for highest column loads.
	Shale	50–2000	to 8	1.3	Shaft* 4–20 Socket 3+	20–80	2–8**	
	Sandstone (Rare in this area)	300–1000	to 8	1.3	Shaft* to 50 Socket 1+	40–70	4–7**	
	Glacial till—very stiff to hard highly plastic clay	100–500	6 (mach.) 6 (hand)	2	Shaft *5–10 Socket 3+	8–16	1.5–3**	Preconsolidation stress may be critical factor. Slickensides and fissuring make shear strength difficult to evaluate.

* "Shaft" is in overburden above rock, "socket" in rock.
** Shear values are not used in combination with end-bearing

265

Locale and source of information	Bearing material	Column load range, kips	Dia. range, ft Bell	Dia. range, ft Shaft or socket	Shaft or socket length/diameter range	End-bearing range, ksf	Sidewall shear range, ksf	Remarks
Los Angeles Area	Dense sand	....	....	....	....	5–15	0.8–15	
Information from:	Pico formation (poorly indurated siltstone to claystone)	....	....	....	....	10–20	0.5–2	See note below.
Woodward-McNeill Associates, Inc., Los Angeles								

Except for the Pico formation, which underlies part of downtown Los Angeles, most of the foundation materials in the area are alluvial soils, variable in composition and condition. Design parameters are generally determined from shearing strength tests. Foundations are usually either straight-shaft "friction" piers, designed for sidewall shear support only, or belled piers with bearing pressures determined from the soil tests, penetration tests, and bearing capacity formulas. Assignment of both sidewall shear support and end-bearing requires a variance from the Building Department.

Locale and source of information	Bearing material	Column load range, kips	Dia. range, ft Bell	Dia. range, ft Shaft or socket	Shaft or socket length/diameter range	End-bearing range, ksf	Sidewall shear range, ksf	Remarks
Omaha, Nebraska Area	Glacial till (Kansas or Nebraskan)	300–2500	To 14	2–3.5	5–10*	10–15	1.5–2	Sidewall shear values for use alone.
	Dakota sandstone (Rare in Omaha area)	250+	3–6			30	....	Data based on one project in Omaha.
Information from:	Loess	Very small to 300	to 10	N.A.		4–6	N.A.	Belled piers used on the desiccated surface of Loveland loess for moderate loads.
Woodward-McMaster & Associates, Omaha	Shale	500–2000	6–20	2.5–4	To 5*	20–40	2–10	Sidewall shear used alone. Experience from Des Moines, Iowa.
	Limestone	Up to 4500	3–6		N.A.	120	Not used	"Dental work" required to prepare bearing surface.

* Shaft length in bearing formation

Locale and source of information	Bearing material	Column load range, kips	Dia. range, ft Bell	Dia. range, ft Shaft or socket	Shaft or socket length/diameter range	End-bearing range, ksf	Sidewall shear range, ksf	Remarks
Philadelphia & Vicinity; Maryland; New Jersey	Very dense decomposed schist/gneiss (Wissahickon formation)	1000–3000	....	3–14	Average 15	25–40*	3–5*	Somewhat discolored brown to gray-brown, more micaceous. N > 60/6″
	Wissahickon schist (gneiss, intact but altered)	2000–4000	....	3–6	Average 20	40–60*	5–10* (but not over 1/5 end-bearing)	Discolored in streaks, more gneissic. Core recovery 30–50%
Information from:	Wissahickon unaltered schist/gneiss	3000–5000	....	3–5	Average 25	60–80	....	Gray coloring, fairly massive. Core recovery >50%
Woodward-Gardner & Associates, Inc., Philadelphia, Pa.	Red shale, Newark series, soft to medium-hard	200–800	....	3–12	Average 15	6–10	....	Weathered top of bedrock. PI < 15. Core recovery >50%
	Red shale, Newark series, medium-hard to hard	800–2000	....	3–12	Average 15	10–25	....	Core does not disintegrate on submergence, but will check. Core recovery >80%

* These values are used both alone and in combined end-bearing and sidewall shear.

St. Louis Area	250-2500	4-9	*	Shaft 10-30	30-50** / 40-60	3-6**	N > 100. Affected by water, slickensides, equipment, etc.
Hard to very hard shale or sandstone							
Medium-hard shale or sandstone	100-1000	2.5-7	*	Shaft 5-25	15-30** / 20-40	2-4**	Bearing values of this order can be used if confirmed by probing below pier bottom.
Information from:	500-3000	2.5-5	1.5-3	Socket 3-5	50-100** / 50-200	5-10**	
Fresh limestone							
Woodward-McMaster & Associates, Inc., St. Louis, Mo.	250-2000	4-7.5	1.5-3	Socket 3	15-40** / 20-60	2-5**	
Slight to moderately weathered limestone							
Residual clays, very stiff to hard	100-500	5-7	*	Shaft 5-10	6-12** / 8-15	0.5-2**	

* Belling buckets can usually underream bases in shales and sandstones more economically than drilling sockets.
*** These shear and end-bearing values to be used in combination.

Texas Gulf Coast—Houston & Vicinity	30-500	6-20	3-6	....	3-10	1-2	Sidewall shear values may be used either alone or combined with end-bearing (straight shafts only).
Clays & sandy clays (Beaumont formation)							
Information from: National Soil Services, Inc., Houston							

San Francisco Bay Area	20-40	....	....	....	....	to 0.5	End-bearing not used.
Loose sand							
Dense sand	75-150	....	....	....	....	0.8-1.2	End-bearing not used.
Medium-stiff clays	50-70	....	....	....	....	0.3-0.8	End-bearing not used.
Information from:	50-150	3-8	....	....	2-10	0.6-1.2	Either end-bearing or sidewall shear used alone, but not in combination.
Stiff clays							
Woodward-Lundgren & Associates, Inc., Oakland, Calif.	100-	3-8	....	....	8-20	1.0-5.0	
Soft rock							
Sheared rock	10,000						

Locale and source of information	Bearing material	Column load range, kips	Dia. range, ft		Shaft or socket length/diameter range	End-bearing range, ksf	Sidewall shear range, ksf	Remarks
			Bell	Shaft or socket				
Seattle, Washington & Puget Sound Area Information from:	Hard glacial till (Gravelly, clayey, sandy silt or silty sand)	20–24,000	4–32	1–2.5	2–30	30	0	Very competent soil. Most common bell dia. 12–18 ft. Most common straight shaft 1.5–2 ft dia.
	Hard silty clay (Glacially preconsolidated)	200–5100	4–18	1–2.5	3.5–30	20	0	Clay is fractured and slickensided; difficult machine belling. Large bells require shoring.
	Very dense sand (Glacially overridden)	200–5100	4–18	1–2.5	3.5–30	<20**	**	** For combined end-bearing and shear, use bearing formulas and $\phi = 36°$ to $38°$; and active pressure. End-bearing often used alone.
Shannon & Wilson, Inc., Seattle, Washington	Stiff to very stiff glacio-marine clay (Slightly overconsol.)	20–150	4–8	......	5–20	2–3	1	Shear for use in combination with end-bearing. End-bearing alone or combined.
	Recent alluvium—bedded medium-stiff silts and loose to medium-dense sands	10–50	......	1–2.5	12–30	1–2	0.25–1.0	These generally include cast-in-place concrete piles in recent alluvium with high groundwater, light column loads.
Washington, D.C. & Vicinity Information from:	Weathered mica schist	500–1000	5–10	2.5–5.5	5–10	10–20	1–2* / 2–3	* Shear value for use in combination with end-bearing.
	Altered mica schist	1500–2500	to 15	to 6	10–15	20–40	5–10	
Woodward-Gardner & Associates, Inc., Bethesda, Md.	Unaltered mica schist/ gneiss/granite	......	......	......	......	~80		Too hard to bell or drill with auger. Socket has to be core-drilled.

Areas and Volumes
of Shafts and Underreams

TABLE F-1   Schedule of Shaft Perimeters, Areas and Volumes and Shaft or Bell End-bearing Areas

Shaft diameter		Shaft perimeter		Shaft or bell end area		Shaft volume	
in.	cm	ft	m	ft^2	m^2	yd^3/ft	m^3/m
18	45.7	4.71	1.44	1.77	0.164	0.06	0.16
20	50.8	5.24	1.60	2.18	0.203	0.08	0.20
22	55.9	5.76	1.75	2.64	0.245	0.10	0.24
24	61.0	6.28	1.92	3.14	0.292	0.12	0.29
26	66.0	6.81	2.08	3.69	0.343	0.14	0.34
28	71.1	7.33	2.23	4.28	0.397	0.16	0.40
30	76.2	7.85	2.39	4.91	0.456	0.18	0.46
32	81.3	8.38	2.55	5.58	0.519	0.21	0.52
34	86.4	8.90	2.71	6.30	0.586	0.23	0.59
36	91.4	9.42	2.87	7.07	0.657	0.26	0.66
38	96.5	9.95	3.03	7.88	0.732	0.29	0.73
40	102	10.47	3.19	8.73	0.811	0.32	0.81
42	107	11.00	3.35	9.62	0.894	0.34	0.89
44	112	11.52	3.51	10.56	0.981	0.39	0.98
46	117	12.04	3.67	11.54	1.072	0.43	1.07
48	122	12.57	3.83	12.57	1.167	0.46	1.17
50	127	13.09	3.99	13.64	1.267	0.50	1.27
52	132	13.61	4.15	14.75	1.370	0.55	1.37
54	137	14.14	4.31	15.90	1.478	0.59	1.48
56	142	14.66	4.47	17.10	1.589	0.63	1.59
58	147	15.18	4.63	18.35	1.705	0.68	1.70
60	152	15.71	4.79	19.64	1.824	0.73	1.82
62	158	16.23	4.95	20.97	1.948	0.78	1.95
64	163	16.76	5.11	22.34	2.075	0.83	2.08
66	168	17.28	5.27	23.76	2.207	0.88	2.21
68	173	17.80	5.43	25.22	2.343	0.93	2.34
70	178	18.33	5.59	26.72	2.483	0.99	2.48
72	183	18.85	5.74	28.27	2.627	1.05	2.63
74	188	19.37	5.90	29.87	2.775	1.11	2.78
76	193	19.90	6.06	31.50	2.927	1.17	2.93
78	198	20.42	6.22	33.18	3.083	1.23	3.08
80	203	20.94	6.38	34.91	3.243	1.29	3.24
82	208	21.47	6.54	36.67	3.407	1.36	3.41
84	213	21.99	6.70	38.48	3.575	1.42	3.58

TABLE F-2 Bell or Underream Volumes—Dome Type
(Volume of shaded portion is volume shown on table.)

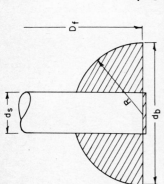

Bell diameter	18″ dia shaft		24″ dia shaft		30″ dia shaft		36″ dia shaft		42″ dia shaft		48″ dia shaft	
	ft³	m³	ft³	m³	ft³	m³	ft³	m³	ft³	m³	ft³	m³
2'-0″ 61 cm	2.5	0.07	4.0	0.11								
2'-6″ 76 cm	4.0	0.11	6.0	0.17								
3'-0″ 91 cm	6.5	0.18	9.0	0.25	6.0	0.17						
3'-6″ 1.07 m	10.0	0.28			8.5	0.22	8.5	0.22				
4'-0″ 1.22 m	15.0	0.42	14.0	0.40	12.5	0.35	11.5	0.33	13.0	0.37	18.0	0.51
4'-6″ 1.37 m			20.5	0.58	18.5	0.52	16.0	0.45	15.0	0.42	21.0	0.59
5'-0″ 1.52 m			28.5	0.81	26.0	0.74	23.5	0.67	20.5	0.58	24.0	0.68
5'-6″ 1.68 m					35.5	1.00	32.5	0.92	29.0	0.82		
6'-0″ 1.83 m					47.0	1.33	43.5	1.23	39.0	1.10	34.5	0.98
6'-6″ 1.98 m					61.5	1.74	57.0	1.61	52.0	1.47	46.5	1.32
7'-0″ 2.13 m					78.0	2.21	72.5	2.05	67.5	1.91	61.0	1.73
7'-6″ 2.29 m					97.5	2.76	92.0	2.60	85.0	2.41	78.0	2.21

TABLE F-2 Bell or Underream Volumes—Dome Type (*Continued*)
(Volume of shaded portion is volume shown on table.)

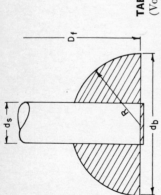

Bell diameter	18" dia shaft		24" dia shaft		30" dia shaft		36" dia shaft		42" dia shaft		48" dia shaft	
	ft³	m³	ft³	m³	ft³	m³	ft³	m³	ft³	m³	ft³	m³
8'-0" 2.44 m							114.0	3.23	106.5	3.01	98.5	2.79
8'-6" 2.59 m							138.5	3.92	131.0	3.71	122.0	3.45
9'-0" 2.74 m							167.0	4.73	158.5	4.49	149.0	4.22
9'-6" 2.90 m									189.5	5.36	179.5	5.08
10'-0" 3.05 m									224.5	6.35	213.0	6.03
10'-6" 3.20 m									263.5	7.46	251.5	7.12
11'-0" 3.35 m											293.5	8.31
11'-6" 3.51 m											340.0	9.62
12'-0" 3.66 m											391.0	11.07

TABLE F-3 Bell or Underream Volumes—45° Type

Outline of bell actually generated · Toe height · 45° · d_s · d_b · L

Bell diameter	18″ dia shaft ft³	18″ dia shaft m³	24″ dia shaft ft³	24″ dia shaft m³	30″ dia shaft ft³	30″ dia shaft m³	36″ dia shaft ft³	36″ dia shaft m³	42″ dia shaft ft³	42″ dia shaft m³	48″ dia shaft ft³	48″ dia shaft m³	54″ dia shaft ft³	54″ dia shaft m³	60″ dia shaft ft³	60″ dia shaft m³
2′ 61 cm	0.9	0.02														
2′6″ 76 cm	2.3	0.06	1.1	0.03												
3′ 91 cm	4.4	0.13	2.9	0.08	1.3	0.04										
3′6″ 1.07 m	7.3	0.21	5.4	0.15	3.5	0.10	1.6	0.04								
4′ 1.22 m	11.1	0.31	8.9	0.25	6.5	0.18	4.1	0.11	1.8	0.05						
4′6″ 1.37 m	15.9	0.45	13.3	0.38	10.5	0.30	7.5	0.21	4.6	0.13	2.1	0.06				
5′ 1.52 m			18.8	0.53	15.5	0.44	12.0	0.34	8.5	0.24	5.2	0.15	2.3	0.07		
5′6″ 1.68 m			25.5	0.72	21.8	0.62	17.7	0.50	13.6	0.39	9.6	0.27	5.8	0.16	2.6	0.07
6′ 1.83 m			33.5	0.95	29.3	0.83	24.7	0.70	20.0	0.57	15.2	0.43	10.6	0.30	6.4	0.18
6′6″ 1.98 m					38.2	1.08	33.1	0.94	27.7	0.78	22.2	0.63	16.7	0.47	11.6	0.33
7′ 2.13 m					48.6	1.38	42.9	1.22	36.9	1.04	30.6	0.87	24.4	0.69	18.3	0.52
7′6″ 2.29 m					60.5	1.71	54.3	1.54	47.6	1.35	40.6	1.15	33.6	0.95	26.6	0.75
8′ 2.44 m							67.4	1.91	60.1	1.70	52.3	1.48	44.4	1.26	36.5	1.03
8′6″ 2.59 m							82.2	2.33	74.2	2.10	65.8	1.86	57.0	1.62	48.2	1.36
9′ 2.74 m							98.9	2.80	90.3	2.56	81.1	2.30	71.5	2.03	61.8	1.75
9′6″ 2.90 m									108	3.07	98.4	2.74	88.0	2.49	77.3	2.19
10′ 3.05 m									128	3.64	118	3.33	106	3.02	94.9	2.69
11′ 3.35 m											163	4.62	150	4.25	137	3.87
12′ 3.66 m											218	6.17	203	5.75	188	5.32
13′ 3.96 m													266	7.54	249	7.05
14′ 4.27 m															322	9.10
15′ 4.57 m															406	11.5

TABLE F-4 Bell or Underream Volumes—30° Type
(Volume of shaded portion is volume shown in table.
Toe height taken as 6 in. for computation of volumes.)

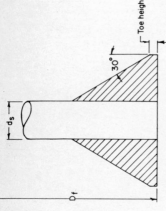

Bell diameter	18″ dia shaft ft³	18″ dia shaft m³	24″ dia shaft ft³	24″ dia shaft m³	30″ dia shaft ft³	30″ dia shaft m³	36″ dia shaft ft³	36″ dia shaft m³	42″ dia shaft ft³	42″ dia shaft m³	48″ dia shaft ft³	48″ dia shaft m³
2′ 61 cm	1.0	0.03										
2′6″ 76 cm	2.8	0.08	1.2	0.04								
3′ 91 cm	5.7	0.16	3.6	0.10	1.5	0.04						
3′6″ 1.07 m	9.8	0.28	7.1	0.20	4.3	0.12	1.8	0.05				
4′ 1.22 m	15.3	0.43	12.0	0.33	8.4	0.24	5.0	0.14	2.1	0.06		
4′6″ 1.37 m	22.4	0.63	18.4	0.52	14.1	0.40	9.8	0.28	5.8	0.16	2.4	0.06
5′ 1.52 m			26.6	0.75	21.5	0.61	16.3	0.46	11.1	0.31	6.5	0.18
5′6″ 1.68 m			36.7	1.04	30.8	0.87	24.6	0.70	18.4	0.52	12.5	0.35
6′ 1.83 m			48.8	1.38	42.2	1.19	35.1	0.99	27.8	0.79	20.6	0.58
6′6″ 1.98 m					55.8	1.58	47.8	1.35	39.3	1.11	30.8	0.87
7′ 2.13 m					71.9	2.01	62.9	1.78	53.3	1.51	43.6	1.23
7′6″ 2.29 m					90.5	2.56	80.5	2.28	70.0	1.98	58.8	1.66

Table 1

		72″ dia shaft		78″ dia shaft		84″ dia shaft	
8′	2.44 m	101	2.86	89.2	2.52	76.9	2.18
8′6″	2.59 m	124	3.52	111	3.14	97.8	2.77
9′	2.74 m	151	4.26	137	3.87	122	3.45
9′6″	2.90 m			165	4.68	149	4.22
10′	3.05 m			197	5.57	180	5.09
10′6″	3.20 m			233	6.60	214	6.06
11′	3.35 m					252	7.13
12′	3.66 m					340	9.61

Table 2

		54″ dia shaft		60″ dia shaft		66″ dia shaft		72″ dia shaft		78″ dia shaft		84″ dia shaft	
5′	1.52 m	2.7	0.08										
5′6″	1.68 m	7.2	0.20	2.9	0.09								
6′	1.83 m	13.8	0.39	8.0	0.23	3.2	0.09						
6′6″	1.98 m	22.7	0.64	15.2	0.43	8.7	0.25	3.5	0.10				
7′	2.13 m	34.0	0.96	24.8	0.70	16.6	0.47	9.4	0.27	3.8	0.11		
8′	2.44 m	64.4	1.82	52.0	1.47	40.2	1.14	29.1	0.82	19.2	0.54	10.9	0.31
9′	2.74 m	106	3.02	90.0	2.57	75.5	2.14	60.5	1.71	46.4	1.31	33.4	0.95
10′	3.05 m	162	4.57	143	4.04	124	3.50	105	2.97	86.6	2.45	69.0	1.95
11′	3.35 m	231	6.54	209	5.92	187	5.28	164	4.64	141	4.00	119	3.36
12′	3.66 m	316	8.95	291	8.24	265	7.50	238	6.75	211	5.97	185	5.24

Conversion Factors, English to Metric (International System of Units)

The following are conversion units for quantities in common use in the aspects of geotechnical engineering discussed in this book—plus a few other related units. For the reader interested in a clear and complete discussion and tabulation of the conversions to and use of the entire International System of Units (SI), the reader is referred to "Metric Practice Guide," ASTM Designation E 380-70, a 33-page booklet obtainable from the American Society for Testing and Materials, 1916 Race Street, Philadelphia, Pennsylvania 19103.

It will be noted that the SI units are, for the most part, the "metric system" or "cgs" units that most readers are familiar with. The principal change is the introduction of the unit of force, the newton (N), taking the place of the kilogram-force (kgf); and the name kilogram by itself designates mass. Note also that the newton is a smaller unit than the kgf; it takes 9.806650 newtons to equal 1 kgf.

Note also that the table that follows presents conversion factors to only three significant figures. This is sufficient for most foundation engineering practice. If greater accuracy is needed, the reader is referred to the "Metric Practice Guide" mentioned above.

To convert from:	To:	Multiply by:
AREA		
foot^2 (sq ft)	meter^2 (m^2)	9.29×10^{-2}
inch^2 (sq in.)	meter^2 (m^2)	6.45×10^{-4}
inch^2 (sq in.)	centimeter^2 (cm^2)	6.45
yard^2 (sq yd)	meter^2 (m^2)	8.36×10^{-1}
TORQUE OR BENDING MOMENT		
pound (force)-foot (lbf-ft)[*]	newton-meter (N-m)	1.36
pound (force)-foot (lbf-ft)	kilogram-force-meter (kgf-m)	1.38×10^{-1}
FORCE		
kilogram-force (kgf)	newton (N)	9.81
kip (kilopound-force)	newton (N)	4.45×10^3
kip (kilopound-force)	kilogram-force (kgf)	4.54×10^2
pound-force (lbf)	newton	4.45
pound-force (lbf)	kilogram-force (kgf)	4.54×10^{-1}
LENGTH		
foot (ft)	meter (m)	3.05×10^{-1}
foot (ft)	centimeter (cm)	3.05×10
inch (in.)	meter (m)	2.54×10^{-2}
inch (in.)	centimeter (cm)	2.54
yard (yd)	meter (m)	9.14×10^{-1}
MASS		
pound-mass (lbm)	kilogram (kg)	4.54×10^{-1}
ton-mass (short, 2000 lbm)	kilogram (kg)	9.07×10^2
kip-mass	kilogram (kg)	4.54×10^2
MASS/VOLUME (DENSITY)		
pound-mass/foot^3[†]	kilogram/meter^3 (kg/m^3)	1.60×10
pound-mass/gallon (U.S.)	kilogram/meter^3 (kg/m^3)	1.20×10^2
pound-mass/cubic yard	kilogram/meter^3 (kg/m^3)	5.93×10^{-1}

[*] Most auger machine manufacturers list their torque ratings as "foot-pounds." This is incorrect. Foot-pounds are units of energy or work, not torque.

[†] Note that measurements of density of any material are measurements of mass per unit volume—not weight (force). A pound and a ton are units of force; a pound-mass is the mass that would weigh one pound on earth at sea level. A kilogram is a unit of mass, convertible to pounds mass.

To convert from: *To:*   *Multiply by:*

PRESSURE OR STRESS (FORCE/AREA)

atmosphere (normal)	newton/meter^2 (N/m^2)	1.01×10^5
foot of water	newton/meter^2 (N/m^2)	2.99×10^3
kilogram-force/centimeter^2 (kgf/cm^2)	newton/meter^2 (N/m^2)	9.81×10^4
kip/foot^2 (ksf)	newton/meter^2 (N/m^2)	4.79×10^5
pound-force/foot^2 (psf)	newton/meter^2 (N/m^2)	4.79×10
pound-force/inch^2 (psi)	newton/meter^2 (N/m^2)	6.89×10^3
pound-force/inch^2 (psi)	kilogram-force/centimeter^2 (kgf/cm^2)	7.03×10^{-2}
ton/foot^2* (tsf)	kilogram-force/centimeter^2 (kgf/cm^2) *	9.77×10^{-1}

* These quantities are usually taken as equivalent in geotechnical engineering practice because field and laboratory measurements are rarely precise enough to warrant making a distinction.

VOLUME

foot^3	meter^3 (m^3)	2.83×10^{-2}
gallon (U.S. liquid)	meter^3 (m^3)	3.79×10^{-3}
gallon (U.S. liquid)	liter	3.79
inch^3 (cu in.)	meter^3 (m^3)	1.64×10^{-5}
liter	meter^3 (m^3)	1.00×10^{-3}
liter	gallon (U.S. liquid)	2.64×10^{-1}
yard^3 (cu yd)	meter^3 (m^3)	7.65×10^{-1}

Index

Index